Vibration Fatigue by Spectral Methods

From Structural Dynamics to Fatigue Damage – Theory and Experiments

Vibration Fatigue by Spectral Methods

From Structural Dynamics to Fatigue Damage – Theory and Experiments

Janko Slavič

Matjaž Mršnik

Martin Česnik

Jaka Javh

Miha Boltežar

ELSEVIER

Elsevier
Radarweg 29, PO Box 211, 1000 AE Amsterdam, Netherlands
The Boulevard, Langford Lane, Kidlington, Oxford OX5 1GB, United Kingdom
50 Hampshire Street, 5th Floor, Cambridge, MA 02139, United States

Library of Congress Cataloging-in-Publication Data
A catalog record for this book is available from the Library of Congress

British Library Cataloguing-in-Publication Data
A catalogue record for this book is available from the British Library

ISBN: 978-0-12-822190-7

For information on all Elsevier publications
visit our website at https://www.elsevier.com/books-and-journals

Publisher: Matthew Deans
Acquisitions Editor: Carrie Bolger
Editorial Project Manager: Emma Hayes
Production Project Manager: Prasanna Kalyanaraman
Designer: Victoria Pearson

Typeset by VTeX

Contents

Nomenclature

Acronyms

AL	$\alpha_{0.75}$ method, first used in Eq. (4.16), page 107
CE	Complex exponential modal identification method, first used in Eq. (3.37), page 89
CSD	Cross spectral density, first used in Eq. (2.43), page 65
DFT	Discrete Fourier transform, first used in Eq. (2.3), page 52
DK	Dirlik method, first used in Eq. (4.28), page 109
DMS	Displacement mode shape, first used in Eq. (1.213), page 44
FFT	Fast Fourier transform, first used in Eq. (2.5), page 52
FRF	Frequency response function, first used in Eq. (1.30), page 9
GM	Gao–Moan method, first used in Eq. (4.17), page 107
IDFT	Inverse discrete Fourier transform, first used in Eq. (2.3), page 52
IRF	Impulse response function, first used in Eq. (1.79), page 19
LSCE	Least squares complex exponential modal identification method, first used in Eq. (3.56), page 92
LSCF	Least squares complex frequency, first used in Eq. (3.62), page 94
LSFD	Least squares frequency domain, first used in Eq. (3.86), page 98
LTI	Linear time invariant system, first used in Eq. (1.57), page 15
MDOF	Multi-degree-of-freedom, first used in Eq. (1.0), page 3
NB	Narrowband method, first used in Eq. (4.10), page 106
PbP	Projection-by-projection, first used in Eq. (5.34), page 125
PDF	Probability density function, first used in Eq. (2.23), page 60
PSD	Power spectral density, first used in Eq. (2.37), page 64
RC	Range-count counting method, first used in Eq. (4.37), page 111
RMS	Root mean square value, first used in Eq. (3.13), page 80
RMS	Root mean square, first used in Eq. (5.28), page 123
SDOF	Single-degree-of-freedom, first used in Eq. (1.0), page 3
SIMO	Single-input multiple-output, first used in Eq. (3.37), page 88
SISO	Single-input single-output, first used in Eq. (3.13), page 81
SMS	Strain mode shape, first used in Eq. (1.213), page 44
SsMS	Stress mode shape, first used in Eq. (1.221), page 46
TB	Tovo–Benasciutti method, first used in Eq. (4.41), page 112
TB1	1st Tovo–Benasciutti method, first used in Eq. (4.43), page 112
TB2	2nd Tovo–Benasciutti method, first used in Eq. (4.44), page 112
WL	Wirsching–Light method, first used in Eq. (4.12), page 106
WL	Wirsching–Light method, first used in Eq. (4.15), page 106
ZB	Zhao–Baker method, first used in Eq. (4.36), page 111

Roman symbols

A	State space matrix, first used in Eq. (1.146), page 30

A	Accelerance, first used in Eq. (1.35), page 10
$\mathbf{a}$	Vector of the numerator polynomial coefficients, first used in Eq. (3.70), page 95
$\mathbf{B}$	State space matrix, first used in Eq. (1.146), page 30
b	The slope of the log-log S-N curve, first used in Eq. (4.2), page 100
b	Weighting parameter in the Tovo–Benasciutti damage intensity, first used in Eq. (4.43), page 112
$\mathbf{b}$	Vector of the denumerator polynomial coefficients, first used in Eq. (3.70), page 95
$\mathbf{C}$	Elasticity matrix, first used in Eq. (1.208), page 43
$\mathbf{C}$	Stress covariance matrix, first used in Eq. (5.18), page 120
$\mathbf{C}$	Viscous damping matrix, first used in Eq. (1.101), page 24
C	Coefficient of the S-N curve, first used in Eq. (4.1), page 100
C	Crest factor, first used in Eq. (3.12), page 80
c	Damping, first used in Eq. (1.1), page 4
c_c	Critical damping, first used in Eq. (1.10), page 6
$\mathbf{C}_{cs}$	Transformation matrix in the Carpinteri and Spagnoli stress criterion, first used in Eq. (5.30), page 124
c_{eq}	Equivalent viscous damping, first used in Eq. (1.50), page 13
$c_{eq,s}$	Equivalent viscous damping for structural damping, first used in Eq. (1.53), page 14
c_n	Fourier (complex) coefficients, first used in Eq. (1.63), page 15
C	Covariance function, first used in Eq. (2.28), page 61
$\mathbf{D}$	Dynamic matrix, first used in Eq. (1.110), page 26
$\mathbf{D}$	Hysteretic damping matrix, first used in Eq. (1.153), page 31
d	Damage intensity/rate, first used in Eq. (4.8), page 104
D	Damage, first used in Eq. (4.6), page 102
$\mathbf{e}$	Strain tensor (short form), first used in Eq. (1.198), page 42
E	Young's modulus of elasticity, first used in Eq. (1.207), page 43
e	Error, first used in Eq. (3.65), page 95
$\mathbf{F}$	Load amplitude vector (in general complex), first used in Eq. (1.163), page 33
$\mathbf{f}$	Load vector, first used in Eq. (1.101), page 24
F	Amplitude of harmonic excitation, first used in Eq. (1.18), page 7
f	Bending stress, first used in Eq. (5.19), page 121
f	Frequency in Hz, first used in Eq. (1.73), page 18
$f_1(t)$	Unit-impulse force, first used in Eq. (1.76), page 18
$F_1(\omega)$	Unit-impulse force in frequency domain, first used in Eq. (1.87), page 20
F_r	Modal force for the rth mode, first used in Eq. (1.165), page 33
$\mathbf{F}(\omega)$	Excitation vector in the frequency domain, first used in Eq. (1.174), page 35
f_e	End frequency in a sweep sine excitation, first used in Eq. (3.6), page 79
$F(\omega)$	Force in (angular) frequency domain, first used in Eq. (1.71), page 17
f_s	Start frequency in a sweep sine excitation, first used in Eq. (3.6), page 79
$f_s(t)$	Sampling frequency or sampling rate, first used in Eq. (2.19), page 54
$f_{s,d}$	Structural/hysteretic damping force, first used in Eq. (1.51), page 13
$f(t)$	Force in time domain, first used in Eq. (1.1), page 4
$f_{v,d}$	Viscous damping force, first used in Eq. (1.47), page 13
G	Shear modulus of elasticity, first used in Eq. (1.207), page 43
G_{xx}	Power (also auto) spectral density (one-sided), first used in Eq. (2.44), page 65
h	Structural/hysteretic damping parameter, first used in Eq. (1.51), page 13
$h(t)$	Unit impulse response function, first used in Eq. (1.79), page 19
H_0	Nondimensional frequency response function, first used in Eq. (1.30), page 9
$\hat{H}_1$	H_1 frequency response estimator, first used in Eq. (3.20), page 83
$\hat{H}_2$	H_2 frequency response estimator, first used in Eq. (3.25), page 84
$\mathbf{h}$	Matrix of time-domain responses in modal identification methods, first used in Eq. (3.55), page 91

$\mathbf{H}_{e_i f}$	Strain frequency response function from the excitation force to strain component e_i, first used in Eq. (1.215), page 45
$\mathbf{H}_{sf}$	Stress frequency response function vector, first used in Eq. (5.14), page 119
$\widetilde{\mathbf{H}}_{sf}$	Reduced stress frequency response function vector, first used in Eq. (6.4), page 128
$\mathbf{H}_{s_i f}$	Stress frequency response function from the excitation force to strain component s_i, first used in Eq. (1.222), page 46
$\widetilde{\mathbf{H}}_{s_i f}$	Reduced stress frequency response function from the excitation force to strain component s_i, first used in Eq. (6.2), page 128
$\mathbf{h}'$	Vector of time-domain responses in modal identification methods, first used in Eq. (3.55), page 91
i	Imaginary unit, first used in Eq. (1.14), page 6
$\mathbf{K}$	Stiffness matrix, first used in Eq. (1.101), page 24
k	Exponent of the S-N curve, first used in Eq. (4.1), page 100
k	Stiffness, first used in Eq. (1.1), page 4
k_r	Modal or generalized stiffness r, first used in Eq. (1.119), page 27
k_u	Kurtosis, first used in Eq. (7.3), page 136
L	Number of response locations, first used in Eq. (3.30), page 85
$\mathbf{M}$	Mass matrix, first used in Eq. (1.101), page 24
m	Mass, first used in Eq. (1.1), page 4
$m(t)$	Measurement noise at the excitation side, first used in Eq. (3.15), page 82
M_j	jth central moment for a discrete series, first used in Eq. (7.2), page 136
m_r	Modal or generalized mass r, first used in Eq. (1.119), page 27
${}_r\mathbf{m}_i$	Spectral moment vector i of mode r, first used in Eq. (6.11), page 129
$\widetilde{\mathbf{m}}_i$	Approximated reduced spectral moment vector, first used in Eq. (6.13), page 130
$\mathbf{M}(\omega)$	Vector of measurement noise at the excitation, first used in Eq. (3.30), page 85
m_i	ith spectral moment, first used in Eq. (2.47), page 66
N	Degrees-of-freedom, first used in Eq. (1.183), page 38
n	Number of cycles at selected level, first used in Eq. (4.6), page 102
$n(t)$	Measurement noise at the response side, first used in Eq. (3.15), page 82
$\mathbf{N}(\omega)$	Vector of measurement noise at the response, first used in Eq. (3.30), page 85
N	Number of cycles-to-failure at selected level, first used in Eq. (4.6), page 102
P	Number of excitation locations, first used in Eq. (3.29), page 85
P_n	Sum of relative powers up to the discrete frequency n, first used in Eq. (3.12), page 80
$p_\mathrm{p}(a)$	Peak amplitude probability density function, first used in Eq. (2.63), page 70
$p(x)$	Probability density function, first used in Eq. (2.24), page 60
q	Shear stress, first used in Eq. (5.5), page 116
Q_r	Modal participation or also scaling factor, first used in Eq. (1.161), page 32
R	Modal constant, first used in Eq. (1.45), page 12
$r(t)$	The shaker excitation signal, first used in Eq. (3.0), page 77
${}_rR_{jk}$	Modal constant related to mode r and excitation k, response j, first used in Eq. (1.169), page 34
R_{ff}	Autocorrelation function of the excitation, first used in Eq. (2.73), page 73
r_pp	Peak-to-peak value, first used in Eq. (3.12), page 80
${}_r\mathbf{R}^{e_i}$	Strain modal constants matrix of component e_i of the rth mode, first used in Eq. (1.217), page 45
${}_r\mathbf{R}^{s_i}$	Stress modal constants matrix of component s_i of the rth mode, first used in Eq. (1.223), page 47
R_L	Lower residual, first used in Eq. (1.230), page 48
${}_r\mathbf{R}$	Modal constants matrix of the mode r, first used in Eq. (1.178), page 36
r_rms	RMS value of r, first used in Eq. (3.12), page 80
R_U	Upper residual, first used in Eq. (1.232), page 48
R_{xx}	Autocorrelation function, first used in Eq. (2.30), page 62
R_{xy}	Cross-correlation function, first used in Eq. (2.30), page 62

$\mathbf{s}$	Stress tensor (short form), first used in Eq. (1.204), page 43
s	Stress tensor component, first used in Eq. (5.9), page 119
s_1	The applied stress amplitude at one-cycle life, first used in Eq. (4.2), page 100
S_{xx}	Power spectral density, first used in Eq. (2.43), page 65
S_{xy}	Cross spectral density, first used in Eq. (2.43), page 65
s_a	Stress amplitude, first used in Eq. (4.4), page 101
$\mathbf{S}_{FF}$	Excitation spectral density matrix, first used in Eq. (3.33), page 86
$\mathbf{S}_{XF}$	Excitation/response spectral density matrix, first used in Eq. (3.33), page 86
$\mathbf{S}_{XX}$	Response spectral density matrix, first used in Eq. (3.33), page 86
$\hat{\mathbf{S}}_{FF}$	Estimator of the excitation spectral density matrix, first used in Eq. (3.33), page 86
$\hat{\mathbf{S}}_{XF}$	Estimator of the excitation/response spectral density matrix, first used in Eq. (3.33), page 86
$\hat{\mathbf{S}}_{XX}$	Estimator of the response spectral density matrix, first used in Eq. (3.33), page 86
$\mathbf{s}$	Stress tensor, first used in Eq. (5.9), page 119
$\mathbf{S}_c$	Equivalent von Mises stress criterion in the frequency domain, first used in Eq. (5.29), page 123
$\mathbf{S}'_{ss}$	Carpinteri and Spagnoli stress criterion in the frequency domain, first used in Eq. (5.30), page 124
$\widetilde{\mathbf{S}}'_c$	Approximated reduced von Mises uniaxial equivalent, first used in Eq. (6.10), page 129
$\mathbf{S}_\mathrm{eq}$	Equivalent stress, first used in Eq. (5.20), page 121
s_eq	Effective cycle amplitude, first used in Eq. (4.4), page 101
S_{ff}	PSD of force excitation, first used in Eq. (5.12), page 119
s_∞	Stress endurance limit, first used in Eq. (4.5), page 101
s_k	Skewness, first used in Eq. (7.3), page 136
s_m	Ultimate tensile strength, first used in Eq. (4.4), page 101
s_r	Stress range, first used in Eq. (4.4), page 101
$\mathbf{S}_{s_i s_i}$	PSD of random stress response for component s_i, first used in Eq. (5.12), page 119
$\mathbf{S}_{ss}$	Cross-correlation stress matrix, first used in Eq. (5.15), page 120
$\widetilde{\mathbf{S}}'_{ss}$	Approximated reduced cross-correlation stress matrix, first used in Eq. (6.8), page 129
$\vec{s}$	Vector in the critical plane, first used in Eq. (5.19), page 121
S_{xx}	Power (also auto) spectral density (two-sided), first used in Eq. (2.39), page 65
S_{xy}	Cross spectral density (two-sided), first used in Eq. (2.39), page 65
S_u	Effective cycle amplitude, first used in Eq. (4.4), page 101
T	Time period, first used in Eq. (1.62), page 15
T	Time to failure, first used in Eq. (4.8), page 104
t	Time, first used in Eq. (1.1), page 4
t	Torsion fatigue limit, first used in Eq. (5.5), page 116
$\mathbf{U}$	Vector of state space complex amplitudes, first used in Eq. (1.147), page 30
$\mathbf{u}$	State vector, first used in Eq. (1.144), page 30
U	Related to $U+1$ time-steps in discrete impulse response function, first used in Eq. (3.50), page 90
$u(t)$	True excitation of a linear SISO system, first used in Eq. (3.15), page 82
$v(t)$	True response of a linear SISO system, first used in Eq. (3.15), page 82
$\mathbf{U}(\omega)$	Vector of true excitation of a linear MIMO system, first used in Eq. (3.30), page 85
$\mathbf{V}$	Compliance matrix, first used in Eq. (1.211), page 44
$\mathbf{V}(\omega)$	Vector of true response of a linear MIMO system, first used in Eq. (3.30), page 85
$W(\omega)$	Weighting function in modal identification methods, first used in Eq. (3.65), page 95
$W_{\mathrm{DM},1}$	Work per harmonic cycle of an arbitrary damping model (DM), first used in Eq. (1.50), page 13
$W_{\mathrm{v,d},1}$	Structural damping work per harmonic cycle, first used in Eq. (1.52), page 14
$W_{\mathrm{v,d}}$	Viscous damping work, first used in Eq. (1.48), page 13
$W_{\mathrm{v,d},1}$	Viscous damping work per harmonic cycle, first used in Eq. (1.49), page 13
$\mathbf{X}$	Vector of displacement amplitudes, first used in Eq. (1.104), page 25

X	Displacement amplitude, first used in Eq. (1.4), page 5
x	Displacement, first used in Eq. (1.1), page 4
$\tilde{X}$	One-sided Fourier transform of a real-valued $x(t)$, first used in Eq. (2.16), page 53
$\mathbf{X}(\omega)$	Response vector in the frequency domain, first used in Eq. (1.174), page 35
$\mathbf{X}_{e_i}$	ith component of the amplitude of strain steady-state response, first used in Eq. (1.215), page 45
x_{h}	Homogeneous solution, first used in Eq. (1.2), page 4
X_k	Spectral component, first used in Eq. (2.3), page 52
x_{p}	Particular solution, first used in Eq. (1.2), page 4
$\tilde{\mathbf{\Psi}}_r$	State-space eigenvector, first used in Eq. (1.149), page 30
$\mathbf{X}_{s_i}$	ith component of the amplitude of stress steady-state response, first used in Eq. (1.222), page 46
$x_s(t)$	Discrete signal, first used in Eq. (2.19), page 54
Y	Mobility, first used in Eq. (1.35), page 10

Greek symbols

η	Damping loss factor, first used in Eq. (1.54), page 14
α	Mass factor in proportional viscous damping, first used in Eq. (1.137), page 29
α	Receptance, the (dimensional) frequency response function, first used in Eq. (1.29), page 9
$\boldsymbol{\alpha}$	Receptance transfer matrix, first used in Eq. (1.175), page 35
$\hat{\alpha}_1$	Receptance α (H_1) frequency response estimator, first used in Eq. (3.20), page 83
$\hat{\alpha}_2$	Receptance α (H_2) frequency response estimator, first used in Eq. (3.25), page 84
α_i	Narrowband/broadband estimator, first used in Eq. (2.49), page 67
α_L	Lower residuals of the receptance, first used in Eq. (1.225), page 47
α_{jk}	Frequency transfer function, j-output/response, k-input/excitation, first used in Eq. (1.169), page 34
α_U	Upper residuals of the receptance, first used in Eq. (1.225), page 47
β	Stiffness factor in proportional viscous damping, first used in Eq. (1.137), page 29
$\boldsymbol{\beta}$	Vector of polynomial coefficients in the in modal identification methods, first used in Eq. (3.55), page 91
β_i	Narrowband/broadband estimator of derivatives, first used in Eq. (2.52), page 67
Δf	Frequency resolution, first used in Eq. (2.4), page 52
$\Delta I_1(t)$	Unit change of the linear momentum, first used in Eq. (1.77), page 18
Δt	Sampling time, first used in Eq. (2.3), page 52
ϵ	Bandwidth estimator, first used in Eq. (2.50), page 67
ϵ	Stiffness factor in proportional hysteretic damping, first used in Eq. (1.154), page 31
ϵ_{V}	Vanmarcke bandwidth estimator., first used in Eq. (2.51), page 67
ζ_r	Modal damping for the rth mode, first used in Eq. (1.166), page 33
$\vec{\eta}$	Vector normal to the critical plane, first used in Eq. (5.19), page 121
Γ	Euler gamma function, first used in Eq. (4.11), page 106
γ	Nonstationarity index, first used in Eq. (7.6), page 137
γ_{fx}^2	Coherence function, first used in Eq. (2.79), page 74
γ_r	Normalization scaling factor for mode r, first used in Eq. (1.121), page 27
j	Response/output location index, first used in Eq. (1.175), page 35
k	Excitation/input location index, first used in Eq. (1.175), page 35
λ	Eigenvalue, first used in Eq. (1.4), page 5
λ_r	rth eigenvalue, first used in Eq. (1.150), page 31
μ	Mean value, first used in Eq. (2.26), page 61
μ_f	Mean value of the excitation, first used in Eq. (2.70), page 72
μ_r	Mean value of a run r, first used in Eq. (7.4), page 137
∇	Nabla-Hamiltonian operator, first used in Eq. (1.200), page 42
ν	Mass factor in proportional hysteretic damping, first used in Eq. (1.154), page 31

ν	Poisson's ratio, first used in Eq. (1.207), page 43
ν_0^+	Frequency of positive slope zero level crossing, first used in Eq. (2.62), page 69
ν_a^+	Frequency of positive-slope level-a-crossing, first used in Eq. (2.57), page 69
$\nu_a^\pm$	Frequency of positive- or negative-slope level-a-crossing, first used in Eq. (2.61), page 69
ν_p	Peak frequency, first used in Eq. (2.66), page 71
ω	Angular frequency, first used in Eq. (1.18), page 7
ω_r	Natural frequency of the rth mode, first used in Eq. (1.166), page 33
ω_0	Undamped angular natural frequency, first used in Eq. (1.9), page 6
ω_d	Damped angular natural frequency, first used in Eq. (1.14), page 6
ω_p	Fundamental angular frequency of a periodic function, first used in Eq. (1.63), page 15
Ω_n	Discrete angular frequency, first used in Eq. (3.64), page 94
$\boldsymbol{\Phi}$	Normalized modal matrix, first used in Eq. (1.128), page 28
$\boldsymbol{\Phi}_r$	Normalized eigenvector, first used in Eq. (1.121), page 27
ϕ	Elements of the mode shape $\boldsymbol{\phi}$, first used in Eq. (1.217), page 45
ϕ^{e_i}	Elements of the strain mode shape $\boldsymbol{\phi}^{e_i}$ of component e_i, first used in Eq. (1.217), page 45
$\boldsymbol{\phi}^{s_i}$	The stress mode shape of the component s_i, first used in Eq. (1.222), page 46
$\boldsymbol{\Phi}_r^{s_i}$	Mass normalized stress mode shape for the component i, first used in Eq. (1.221), page 46
$\boldsymbol{\Phi}_r^{\varepsilon}$	Mass normalized strain mode shape, first used in Eq. (1.214), page 44
$\Phi(x)$	Standard normal PDF ($\mu = 0$ and $\sigma_x = 1$), first used in Eq. (2.67), page 71
$\boldsymbol{\Psi}_r$	Eigenvector/mode shape of the rth mode, first used in Eq. (1.108), page 25
ρ_WL	Wirsching–Light correction factor, first used in Eq. (4.12), page 106
$\boldsymbol{\sigma}$	Stress tensor (short form), first used in Eq. (1.202), page 42
σ	Standard deviation, first used in Eq. (2.26), page 61
σ_af	Bending stress, first used in Eq. (5.5), page 116
$\sigma_{\dot{x}}^2$	Variance of the first time derivative, first used in Eq. (2.53), page 68
$\sigma_{\ddot{x}}^2$	Variance of the second time derivative, first used in Eq. (2.53), page 68
$\overline{\boldsymbol{\sigma}}$	Stress tensor, first used in Eq. (1.202), page 42
σ_r	Variance of a run r, first used in Eq. (7.4), page 137
τ_af	Bending fatigue limit, first used in Eq. (5.5), page 116
$\boldsymbol{\varepsilon}$	Strain tensor (short form), first used in Eq. (1.196), page 41
$\overline{\boldsymbol{\varepsilon}}$	Strain tensor, first used in Eq. (1.196), page 41
φ	Phase delay, first used in Eq. (1.43), page 12
$\varphi(t)$	Temporal phase of sine sweep excitation, first used in Eq. (3.5), page 79
$\varrho_r(\omega)$	Base function, first used in Eq. (3.63), page 94
ζ	Damping ratio, first used in Eq. (1.11), page 6

List of operators

$(\cdot)^*$	Complex conjugate, first used in Eq. (1.66), page 16
$(\cdot)^{*\mathrm{T}}$	Transpose of complex conjugated matrix, first used in Eq. (1.159), page 32
$(\cdot)^{*-\mathrm{T}}$	Transpose of inverse complex conjugate matrix, first used in Eq. (1.187), page 39
$\mathrm{D}\{\cdot\}$	Linear spatial differential operator, first used in Eq. (1.199), page 42
$E[\cdot]$	Statistical expectation, first used in Eq. (2.27), page 61
$\mathcal{F}\{\cdot\}$	Forward Fourier transform, first used in Eq. (1.70), page 17
$\mathcal{F}^{-1}\{\cdot\}$	Inverse Fourier transform, first used in Eq. (1.70), page 17
$\Im\{\cdot\}$	Imaginary part of, first used in Eq. (1.36), page 11
$\Pr\{\cdot\}$	Probability, first used in Eq. (2.55), page 68
$\Re\{\cdot\}$	Real part of, first used in Eq. (1.36), page 11
$\S\{\cdot\}$	Linear operator from infinitesimal displacement field to infinitesimal stress tensor, first used in Eq. (1.220), page 46
$(\cdot)^\mathrm{T}$	Transpose, first used in Eq. (1.112), page 26
$(\cdot)_\mathrm{h}$	Homogeneous part of the solution of the differential equation, first used in Eq. (1.2), page 4

$(\cdot)_p$	Particular part of the solution of the differential equation, first used in Eq. (1.2), page 4
$\angle\{\cdot\}$	Angle/phase of a complex number, first used in Eq. (1.38), page 11

List of selected indexes

j	Response location, first used in Eq. (1.189), page 39
k	Excitation location, first used in Eq. (1.189), page 39
r	Eigenvalue or modal index, first used in Eq. (1.189), page 39

Introduction

If a structure is excited close to a natural frequency, then the response of the structure is increased even if the excitation force is relatively small. A real structure typically has more than one natural frequency excited, and the excitation is typically present in more directions, leading to complex stress/strain loads inside the structure. The loads complexity is in the time, frequency, and spatial domains.

Vibration fatigue relates structural dynamics with the material fatigue. Structural dynamics provides the theoretical background on the dynamic response of vibrating structures. The structural dynamics theory deals with multibody dynamical systems; it relates the excitation (e.g., force or kinematic) of those systems to the response (e.g., displacement or stress). The relations from the excitation to response are defined in the frequency domain. The structural dynamics models are validated using the experimental modal analysis.
The structural dynamics theory in this book is extended toward the material fatigue. Initially, the uniaxial stress response, calculated in the frequency domain, is discussed. The basic assumptions used are: the high-cycle fatigue, S-N curve material data, and the Palmgren–Miner hypothesis of damage accumulation. To obtain the cycle-amplitude probability density estimation, the spectral moment approach is used. If multiaxial loads are researched, then the multiaxial criteria is used to obtain the equivalent uniaxial stress response.

This book is organized in two parts. Part I gives the theoretical background, and Part II gives the selected experimental research. We now give a short overview of chapters.

Ch. 1 gives the theoretical background on the *structural dynamics*.

This chapter is organized in two sections. The first section gives the theoretical background on the single-degree-of-freedom systems, whereas the second section is on the multi-degree-of-freedom systems. Initially, damping-free systems are discussed, but then the damping in the form of the viscous and hysteretic model is presented. Free and forced responses are discussed. The theory for harmonic, periodic, and general forcing excitation is presented. The strain/stress response, later important in fatigue estimation, is related to the theory of structural dynamics.

In real applications the dynamically loaded structures are attached to the supporting vibrating structure (e.g., an alternator to the engine), whereas in the laboratory the dynamical loads are usually applied using an (electrodynamic) shaker. Without proper care, the real application and laboratory replication can lead to significantly different dynamical systems, and therefore this chapter also discusses the details of base-excited structures and the difference of the dynamic excitation and the kinematic excitation.
Finally, incomplete models are discussed as flexible structures have, theoretically, an infinite number of natural frequencies, whereas the experimental/numerical models are limited with regard to the number of natural frequencies.

Ch. 2 gives the theoretical background on *signal processing* and is divided into two sections, the deterministic process and the random process.

The deterministic process section discusses the Fourier transform and the discrete Fourier transform. It further discusses the properties of the Fourier transform, which are especially important when handling measured data. The most important practical aspects with regard to data acquisition and signal processing are also discussed (e.g., sampling frequency, aliasing, window functions).

The random process section presents a generalized framework for dealing with data in the frequency domain. A special attention is given to the Gaussian random process as most of random processes can be considered Gaussian. Besides the moments of the amplitude probability density function, also the very important property of ergodicity is discussed. Power spectral density, as the basic tool of describing random data in the frequency domain, is related to spectral moments. Further, the bandwidth of the random process and the peak and level-crossing frequency are discussed. Finally, the response of a linear time-invariant system when randomly excited is discussed.

The deterministic and random process section presents a strong theoretical basis for the experimental modal analysis and for the vibration fatigue discussed in the subsequent chapter.

Ch. 3 gives the theoretical background on the *experimental modal analysis*, which is an important tool for model validation; this chapter is divided into three sections: Types of excitation, Frequency response function estimation, and Modal parameter identification.

Types of excitation in experimental modal analysis and vibration fatigue are very important. Here the impact and (electrodynamic) shaker excitation are presented in detail. From structural point of view, it is important to have the frequency and amplitude of excitation under control in the time and frequency domain. The type of excitation depends on the application.

The frequency response function is theoretically defined already in Ch. 1; however, experimental identification is always contaminated with measurement noise. The noise is present at the input/excitation output/response measurements. This section deals with the measurement and signal processing details, which are important in obtaining a good frequency response function estimator and also have its quality under control.

The final section of Ch. 3 deals with modal parameter estimation methods. The section initially gives theoretical background for simpler methods and then extended toward advanced methods, which are typically used today. Four methods are presented in detail.

Ch. 4 gives the theoretical background on the *uniaxial vibration fatigue*. The goal of this chapter is giving the theoretical background on the uniaxial high-cycle loads typical in vibration fatigue. The uniaxial approach will be given in the subsequent Ch. 5 generalized to the multiaxial case.

In the first section the basics of the time-domain stress approach to material fatigue at harmonic, constant amplitude, loads are presented. The second section extends the theory to variable amplitude loads and ends up with the basics of the probabilistic approach to fatigue estimation. The last section extends the probabilistic approach to the frequency domain (also known under the term spectral (moments) methods). Based on the uniaxial fatigue and the signal processing theory, a list of established frequency domain methods is theoretically presented.

The theory of Ch. 4 is related to the following chapters in the experimental part.

Part II: Ch. 8 Accelerated vibration-fatigue testing presents an experimental approach for the harmonic accelerated vibration testing with concurrent identification of fatigue and structural parameters.

Part II: Ch. 9 Harmonic vs. random excitation presents an experimental approach accelerated vibration fatigue testing with random and harmonic excitation. A comparison of experimental and numerical results is also given.

Part II: Ch. 10 Frequency domain methods for fatigue life estimation experimentally compares the methods presented in Ch. 4.

Ch. 5 extends the uniaxial loads theory to the *multiaxial loads theory* in three sections.

The first section *Time domain basics* discusses the historic origins of multiaxial loads and presents relevant basic prior research and the nomenclature. The second section *Random stress response* relates the multiaxial stress state introduced in Ch. 1 to the random stress response; using the force to the stress-component frequency-response-functions, we introduce a generalization in the form of stress spectral density matrix.

The third section *Frequency-domain multiaxial criteria* discusses four different multiaxial methods: the critical plane criteria, equivalent von Mises stress, Carpinteri–Spagnoli, and projection-by-projection methods.

The theory of Ch. 5 is related to the following chapter in the experimental part:

Part II: Ch. 11 Multiaxial vibration fatigue presents an experimental setup where the Y-sample has been multiaxially excited. The different theoretical methods are compared and discussed on the experimental data.

Ch. 6 discusses the *modal decomposition based vibration fatigue analysis.* Whereas this could be considered a multiaxial criterion, it presents a theoretical generalization of the multiaxial loads. As is discussed in Ch. 1, the dynamic model can be researched using the spatial, modal, or response model. The FEM model presents a spatial model and is defined using the physical coordinates; similarly, the multiaxial methods, discussed in Ch. 5, are also based on the physical coordinates. Using the modal model and therefore the modal coordinates moves the frequency domain damage estimation to the modal space resulting in several advantages (e.g., the analysis of damage contribution of each mode, the reduced model analysis, significant numerical speed-up). As the majority of fatigue damage of vibrating structures occurs at the natural frequencies, the modal decomposition approach looks very promising as it decomposes the damage to particular natural frequencies and corresponding mode shapes.

Ch. 6 has four sections, where the first section discusses the stress response of the reduced model, and the second section discusses the damage intensity estimation based on the modal decomposition. The third section presents a numerical experiment, and the final section is the discussion.

Ch. 7 presents the vibration fatigue at *nonstationary and non-Gaussian excitation*. All previous chapters are based on the assumptions of stationary and Gaussian excitation. However, dynamical systems are frequently nonstationary and non-Gaussian excited, and this chapter is divided into two sections: the first deals with the identification on nonstationary excitation, and the second deals with the identification of the non-Gaussianity.

The theory of this chapter is related to the following chapters in the experimental part:

Part II: Ch. 12 Vibration fatigue for nonstationary and non-Gaussian excitation presents an experiment with the identification of nonstationary and non-Gaussian excitation. Further, the significance of nonstationary vs. non-Gaussian excitation is researched.

Part I

Theoretical background

Chapter 1

Structural dynamics

Structural dynamics researches the dynamics of systems or structures that have usually continuously distributed mass, damping, and stiffness parameters. A continuous system has theoretically infinite number of degrees of freedom. Simple structures (e.g., beams) can be handled analytically, whereas complex structures are typically handled using the finite element method, which has limited number of degrees of freedom. In this chapter, we will initially give the theoretical background on classification and basic terminology and then continue with general background on structural dynamics related to vibration fatigue, in particular:

- Section 1.1: discusses a single-degree-of-freedom system,
- Section 1.2: expands the theory to a general multi-degree-of-freedom system, and finally,
- Section 1.2.5: discusses the strain and stress response of single- and multi-degree-of-freedom systems.

The readers interested in greater detail on structural dynamics are advised to specialized books on mechanical vibrations, for example, [1–4].

Classification

There are many ways mechanical vibrations can be classified. The basic classification is with regard to the number of degrees of freedom. Typically, the theory starts with a *single-degree-of-freedom* (SDOF) system, which is then expanded to a *multi-degree-of-freedom* (MDOF) system.

Further, mechanical vibrations are typically classified into *free* and *forced* vibrations. In case of free vibrations, after initial disturbance (e.g., impact excitation), the system is left to vibrate on its own. In case of forced vibrations the system is excited by an external force. In vibration fatigue, we are primarily interested in forced vibrations, where mechanical systems are excited with a dynamic external force.

The excitation forces can be understood relative to the natural dynamics of the researched system or structure. The force is considered *dynamic* if its frequency range is in the frequency range of the natural dynamics of the researched system or structure. If the force is varying, but the frequency range of the force is significantly below the frequency range of the natural dynamics of the researched system, then the force is considered *quasistatic*. Finally, the force is considered *static* if the force is constant in time. An example of a static force

Vibration Fatigue by Spectral Methods. https://doi.org/10.1016/B978-0-12-822190-7.00008-6

is the force due to the gravity, and an example of a quasistatic force is that applied during a classical tensile test. Examples of dynamic forces are: wave forces applied to offshore structures, earthquake forces applied to tall buildings, and forces due to the unbalanced masses in reciprocating engines applied to structures attached to the engine.

Dynamic excitation forces are further classified into *impact*, *periodic*, and *broadband* random excitation. The impact forces are typically very short in time domain (theoretically, the Dirac delta function) and relatively wide in the frequency domain. Periodic excitation is usually performed using a *harmonic* force, that is, a sinusoidal force excitation of a selected frequency and amplitude; with *sweep-sine excitation*, the frequency (possibly also the amplitude) of harmonic excitation continuously changes with time. For the broadband random excitation, the random distribution is focused on a selected frequency range. With impact excitation, the *transient response* is excited, whereas with harmonic and broadband random excitation, we typically focus on the *steady-state response*.

1.1 Single-degree-of-freedom systems (SDOF)

Fig. 1.1 shows a typical single-degree-of-freedom (SDOF) system comprised of *mass* m, spring with *stiffness* k, and *viscous damping* c.

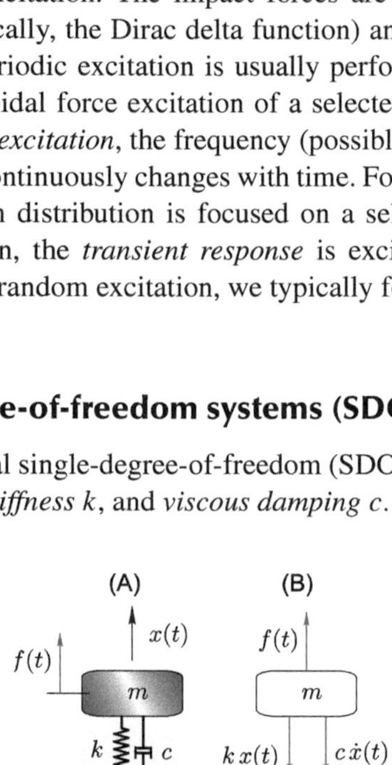

FIGURE 1.1 A single-degree-of-freedom system: (A) the model, (B) the free-body diagram.

The model of an SDOF system can be described with an inhomogeneous second-order differential equation (see free-body diagram in Fig. 1.1B):

$$m\,\ddot{x}(t) + c\,\dot{x}(t) + k\,x(t) = f(t), \tag{1.1}$$

where $f(t)$ is the excitation force. The solution of the inhomogeneous equation is in the form of

$$x(t) = x_{\mathrm{h}}(t) + x_{\mathrm{p}}(t), \tag{1.2}$$

where x_{p} is the particular solution, and x_{h} is the homogeneous solution.

1.1.1 Free-response

Without external force $f(t) = 0$, the homogeneous solution is obtained from

$$m\,\ddot{x}_{\mathrm{h}}(t) + c\,\dot{x}_{\mathrm{h}}(t) + k\,x_{\mathrm{h}}(t) = 0 \tag{1.3}$$

by assuming that the solution $x_{\mathrm{h}}(t)$ is of the form

$$x_{\mathrm{h}}(t) = X_{\mathrm{h}}\,\mathrm{e}^{\lambda\,t}, \tag{1.4}$$

where X_{h} and λ are the unknown (to be determined) displacement amplitude and eigenvalue, respectively. With assumption (1.4), the homogeneous part (1.3) takes the form

$$\lambda^2\,m\,X_{\mathrm{h}}\,\mathrm{e}^{\lambda\,t} + \lambda\,c\,X_{\mathrm{h}}\,\mathrm{e}^{\lambda\,t} + k\,X_{\mathrm{h}}\,\mathrm{e}^{\lambda\,t} = 0. \tag{1.5}$$

After extraction we have

$$\left(\lambda^2\,m + \lambda\,c + k\right)\,X_{\mathrm{h}}\,\mathrm{e}^{\lambda\,t} = 0, \tag{1.6}$$

so that the nontrivial solution ($X_{\mathrm{h}}\,\mathrm{e}^{\lambda\,t} \neq 0$) can be obtained from the characteristic equation

$$\lambda^2\,m + \lambda\,c + k = 0. \tag{1.7}$$

The solution of the characteristic equation (1.7) is

$$\lambda_{1,2} = -\frac{c}{2\,m} \pm \frac{\sqrt{c^2 - 4\,k\,m}}{2\,m}, \tag{1.8}$$

where λ^2 are *eigenvalues* of system (1.1) and are in general complex numbers.

With regard to the relation between the inertial, elastic, and damping forces the solution is:

overdamped: if the damping forces prevail the elastic and inertial forces ($c^2 \geq 4\,k\,m$),

critically damped: if the damping forces equal the elastic and inertial forces ($c^2 = 4\,k\,m$),

underdamped: if the elastic and inertial forces prevail the damping forces ($c^2 \leq 4\,k\,m$).

In case of overdamped and critically damped system, the solution is a real number, and motion is not oscillatory.

In vibration fatigue research, we focus on the underdamped case where the eigenvalues are complex, resulting in oscillatory motion. Due to this reason, in this book, we focus on the underdamped case only.

In case of underdamped systems the complex eigenvalues are conjugated pairs, and the mass m, damping c, and stiffness k (1.3) can be related to the *undamped angular natural frequency* (also *resonant frequency, eigenfrequency*)

$$\omega_0 = \sqrt{\frac{k}{m}}. \tag{1.9}$$

It is important to note that, in linear systems, the prestress does not change the natural frequency.

Further, if *critical damping* is defined as

$$c_c = 2\sqrt{k\,m}, \tag{1.10}$$

then the damping can be defined using the *damping ratio*

$$\zeta = \frac{c}{c_c}. \tag{1.11}$$

With the undamped natural frequency ω_0 and the damping ratio ζ, the homogeneous differential equation (1.3) can be rewritten in the standard form

$$\ddot{x}_h(t) + 2\,\zeta\,\omega_0\,\dot{x}_h(t) + \omega_0^2\,x_h(t) = 0. \tag{1.12}$$

Further, ω_0 and ζ can be inserted into Eq. (1.8):

$$\lambda_{1,2} = -\zeta\,\omega_0 \pm \mathrm{i}\,\omega_d, \tag{1.13}$$

where

$$\omega_d = \omega_0\sqrt{1-\zeta^2} \tag{1.14}$$

is called the *damped angular natural frequency*.

The response of the system is obtained using the eigenvalues from Eq. (1.13) in Eq. (1.4):

$$\begin{aligned} x_h(t) &= X_{h,1}\,e^{\lambda_1 t} + X_{h,2}\,e^{\lambda_2 t} \\ &= e^{-\zeta\,\omega_0 t}\left(X_{h,1}\,e^{+\mathrm{i}\,\omega_d t} + X_{h,2}\,e^{-\mathrm{i}\,\omega_d t}\right), \end{aligned} \tag{1.15}$$

where $X_{h,1}$ and $X_{h,2}$ depend on the initial condition.

Later, it will be clear that the initial conditions where the system starts from the equilibrium with nonzero velocity,

$$\begin{aligned} x(0) = 0 &\rightarrow & X_{h,1} + X_{h,2} &= 0, \\ \dot{x}(0) = \dot{x}_0 &\rightarrow & X_{h,1}\,\lambda_1 + X_{h,2}\,\lambda_2 &= \dot{x}_0, \end{aligned} \tag{1.16}$$

are particularly important. From Eq. (1.16) we obtain the following solution:

$$x_h(t) = \frac{\dot{x}_0}{\omega_d}\,e^{-\zeta\,\omega_0 t}\,\sin(\omega_d\,t). \tag{1.17}$$

The response consists of the decaying envelope $\dot{x}_0/\omega_d \exp(-\zeta\,\omega_0 t)$, which is characterized by the decay rate $\zeta\,\omega_0$ and the oscillating part $\sin(\omega_d t)$; see Fig. 1.2.

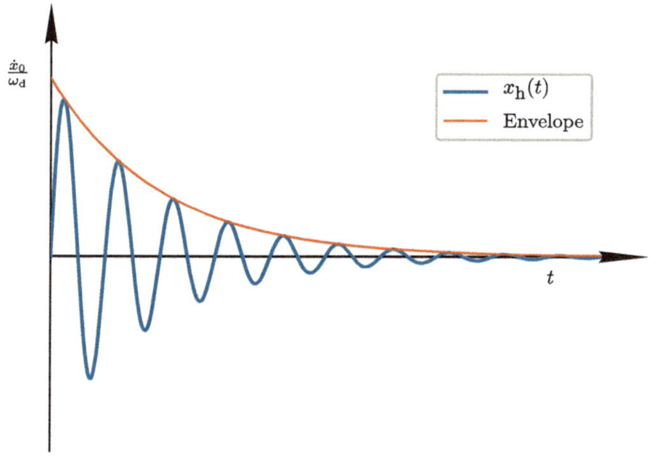

FIGURE 1.2 The free response of an SDOF model.

1.1.2 Harmonic excitation/response

We can now continue with the solution of the equation of motion (1.1) for the case of harmonic excitation:

$$f(t) = F\,e^{i\,\omega t} = F\,(\cos(\omega t) + i\,\sin(\omega t))\,, \tag{1.18}$$

where F is the amplitude of force (in general, a complex number), and ω is the angular frequency of excitation. In Eq. (1.18) the *Euler formula* is used:

$$e^{i\,\omega t} = (\cos(\omega t) + i\,\sin(\omega t))\,. \tag{1.19}$$

In this case the particular solution (1.2) is obtained from

$$m\,\ddot{x}_p(t) + c\,\dot{x}_p(t) + k\,x_p(t) = f(t). \tag{1.20}$$

Assuming a harmonic response

$$x_p(t) = X_p\,e^{i\,\omega t}, \tag{1.21}$$

where X_p is the unknown *complex amplitude* of the particular part (sometimes also called the *phasor*), Eq. (1.20) can be rewritten as

$$-\omega^2\,m\,X_p\,e^{i\,\omega t} + i\,\omega\,c\,X_p\,e^{i\,\omega t} + k\,X_p\,e^{i\,\omega t} = F\,e^{i\,\omega t}. \tag{1.22}$$

Since $e^{i\,\omega t} \neq 0$, Eq. (1.22) simplifies to

$$-\omega^2 m\, X_p + i\,\omega c\, X_p + k\, X_p = F. \tag{1.23}$$

Finally, the ratio between the (response) displacement and (excitation) force amplitudes is

$$\frac{X_p}{F} = \frac{1}{-\omega^2 m + i\,\omega c + k}. \tag{1.24}$$

For two different damping ratios, the frequency response X_p/F is also shown in Fig. 1.3, where the *relative frequency*[1]

$$r = \omega/\omega_0 \tag{1.25}$$

is used.

The particular solution (1.21) can be written also as

$$x_p(t) = \frac{1}{-\omega^2 m + i\,\omega c + k}\, F\, e^{i\,\omega t}. \tag{1.26}$$

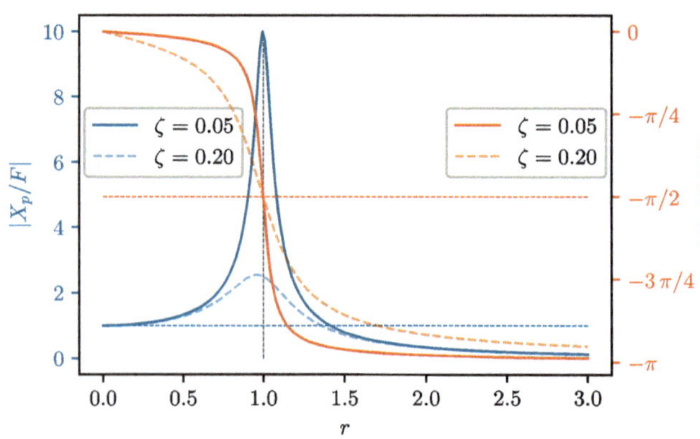

FIGURE 1.3 Amplitude and phase of the frequency response function of an SDOF model.

With Eq. (1.24), the particular solution $x_p(t)$ (1.21) is fully defined and thus also the complete solution of Eq. (1.2).

Transient vs steady-state response

Fig. 1.4 shows the homogeneous $x_h(t)$, the particular $x_p(t)$, and the complete $x(t)$ solution of the harmonically excited SDOF model. From the complete solution (1.2) it is clear that, due to damping, the homogeneous part relatively quickly diminishes and only the particular part remains.

[1] Sometimes *frequency ratio* is used.

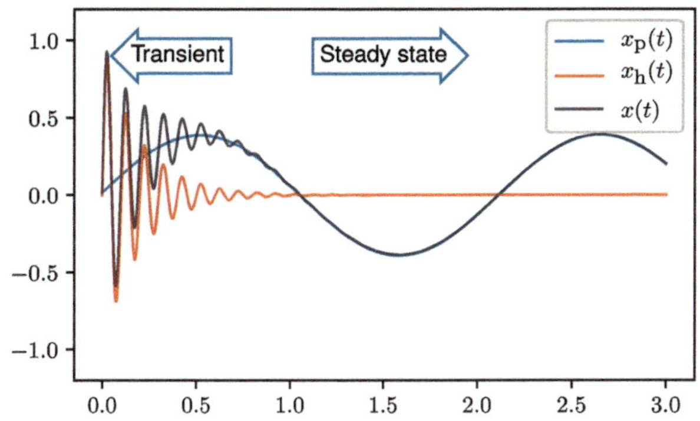

FIGURE 1.4 The superposition of $x_\mathrm{h}(t)$ and $x_\mathrm{p}(t)$ with the transient/steady-state response.

The frequency response function

Initially, we focus on the particular solution given in Eq. (1.24) and rewrite it without the subscript "p":

$$\frac{X}{F} = \frac{1}{-\omega^2 m + \mathrm{i}\,\omega c + k}. \tag{1.27}$$

As Eq. (1.27) presents the frequency response, it is called the *frequency response function* (FRF)[2]:

$$\alpha(\omega) = \frac{1}{-\omega^2 m + \mathrm{i}\,\omega c + k}. \tag{1.28}$$

Using the undamped natural frequency $\omega_0^2 = k/m$ (1.9) and the relation of the damping c via the damping ratio (1.11) to the mass and stiffness (1.10) $c = \zeta\, 2\sqrt{k\, m}$ and finally adding the relative frequency r (1.25), we obtain the frequency response in the standard form

$$\alpha(\omega) = \frac{1}{k} \cdot \frac{1}{1 - r^2 + 2\,\mathrm{i}\,\zeta\, r} \tag{1.29}$$

which is called the *frequency response function* or exactly, the *dimensional frequency response function*. In contrast to $\alpha(\omega)$, we define

$$H_0(\omega) = \frac{X(\omega)}{X_0}, \tag{1.30}$$

which in some books (e.g., [2]) is called the *nondimensional frequency response function*.

[2] In Section 1.1.4, we will show that the same FRF can be obtained from the homogeneous solution.

The *static deflection* X_0 due to the *static force* F is defined as

$$X_0 = \frac{F}{k}. \tag{1.31}$$

Using $F = X_0 k$ with Eq. (1.27) for the harmonic response $X(\omega)$, the nondimensional frequency response function (FRF) can be defined as

$$\begin{aligned} H_0(\omega) &= \frac{X(\omega)}{X_0} \\ &= \frac{k}{-\omega^2 m + \mathrm{i}\,\omega c + k} \\ &= \frac{1}{1 - r^2 + \mathrm{i}\,2\,\zeta\,r}, \end{aligned} \tag{1.32}$$

where $r = \omega/\omega_0$.

The relation between both types of FRF is

$$\alpha(\omega) = \frac{H_0(\omega)}{k}. \tag{1.33}$$

With the FRF, the steady-state response at harmonic excitation (1.26) with angular frequency ω_f can be rewritten as

$$x(t) = \alpha(\omega = \omega_f)\,F\,\mathrm{e}^{\mathrm{i}\,\omega_f\,t}. \tag{1.34}$$

Different forms of FRF

The dimensional FRF makes sense as the static deflection X_0 is hard to measure, and frequently the excitation force and response acceleration (or velocity, or displacement) are measured (1.27). Consequently, different forms of frequency response function are defined as:

$$\begin{aligned} \alpha(\omega) &= \frac{X(\omega)}{F(\omega)}, && \text{receptance,} \\ Y(\omega) &= \frac{\dot{X}(\omega)}{F(\omega)} = \mathrm{i}\,\omega\,\alpha(\omega), && \text{mobility,} \\ A(\omega) &= \frac{\ddot{X}(\omega)}{F(\omega)} = -\omega^2\,\alpha(\omega), && \text{accelerance,} \end{aligned} \tag{1.35}$$

where $\dot{X}(\omega)$ and $\ddot{X}(\omega)$ correspond to the complex amplitudes of velocity and acceleration in the frequency domain, respectively.

In Fig. 1.5 the receptance, mobility, and accelerance of the same frequency response function are compared. The conversions based on derivation/integration in frequency domain[3] between different FRF forms are given in Table 1.1.

[3] See also derivation/integration in Section 2.1.4.

The inverse of accelerance is the *dynamic mass* ($F/\ddot{X}$), the inverse of mobility is the *mechanical impedance* ($F/\dot{X}$), and the inverse of the receptance is the *dynamic stiffness* (F/X).

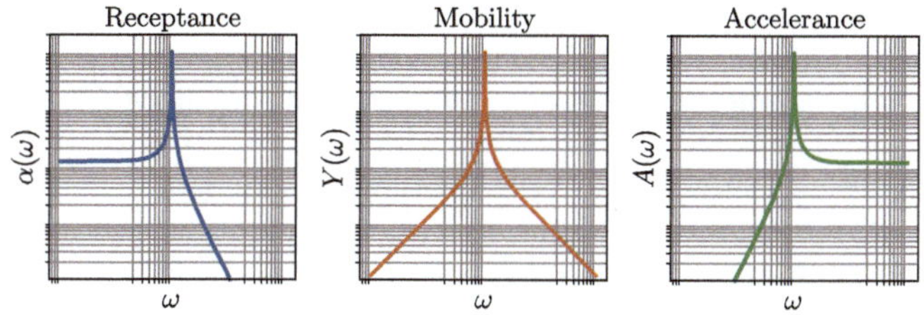

FIGURE 1.5 Receptance, mobility, and accelerance.

TABLE 1.1 A conversion table between different forms of frequency response functions.

To (/)	From (·)		
	α	Y	A
α	1	$1/\mathrm{i}\,\omega$	$-1/\omega^2$
Y	$\mathrm{i}\,\omega$	1	$1/\mathrm{i}\,\omega$
A	$-\omega^2$	$\mathrm{i}\,\omega$	1

Polar representation of the FRF

A complex number C can be represented in the *rectangular form*

$$C = \Re\{C\} + \mathrm{i}\,\Im\{C\} \tag{1.36}$$

or in the *polar form*

$$C = |C|\,\mathrm{e}^{\mathrm{i}\,\angle C}, \tag{1.37}$$

where the amplitude $|C|$ is defined as

$$|C| = \sqrt{\Re\{C\}^2 + \Im\{C\}^2} \tag{1.38}$$

and the phase as

$$\angle C = \arctan\left(\frac{\Im\{C\}}{\Re\{C\}}\right). \tag{1.39}$$

To write the nondimensional FRF (1.32) in the polar form, we initially take into account the quotient form, where the nominator N and denominator D are

$$H_0(\omega) = \frac{N}{D} = \frac{1}{1 - r^2 + \mathrm{i}\,2\,\zeta\,r}. \tag{1.40}$$

For a quotient of two complex numbers, we have [1]

$$\left|\frac{N}{D}\right| = \frac{|N|}{|D|} \quad \text{and} \quad \frac{N}{D} = \frac{|N|\,\mathrm{e}^{\mathrm{i}\,n}}{|D|\,\mathrm{e}^{\mathrm{i}\,d}} = \frac{|N|}{|D|}\,\mathrm{e}^{\mathrm{i}\,(n-d)}, \tag{1.41}$$

where n and d are phase angles. The nominator in (1.40) is a real number with zero phase. The amplitude of the denominator (1.38) therefore is

$$|D| = \sqrt{\left(1-r^2\right)^2 + (2\,\zeta\,r)^2}, \tag{1.42}$$

and the phase (1.39) is

$$\varphi = \angle d = \arctan\left(\frac{2\,\zeta\,r}{1-r^2}\right). \tag{1.43}$$

Instead of $\angle d$, we typically use the symbol φ.

Consequentially, the nondimensional FRF polar form is

$$\begin{aligned} |H_0| &= \frac{1}{\sqrt{\left(1-r^2\right)^2 + (2\,\zeta\,r)^2}}, \\ \angle H_0 &= -\arctan\left(\frac{2\,\zeta\,r}{1-r^2}\right). \end{aligned} \tag{1.44}$$

Partial fraction expansion

Later, the receptance (1.35) will also be used in the *partial-fraction expansion form* (also, *pole-residual* format):

$$\alpha(\omega) = \frac{R}{\mathrm{i}\,\omega - \lambda} + \frac{R^*}{\mathrm{i}\,\omega - \lambda^*}, \tag{1.45}$$

where λ are the complex conjugate solutions defined by Eq. (1.13): $\lambda = \lambda_1 = \lambda_2^*$. The separate parts indicate complex conjugate separated modes; in the SDOF case the separated parts base on the complex conjugate solutions of the characteristic Eq. (1.13), and the nominators contain the *modal constant/pole residue*

$$R = \frac{1}{2\,\mathrm{i}\,m\,\omega_\mathrm{d}}. \tag{1.46}$$

In case of *lightly damped systems* ($\zeta \ll 1$), it is common to use $\omega_\mathrm{d} \approx \omega_0$ in Eq. (1.46); see also Eq. (1.14).

Equivalent viscous and structural/hysteretic damping

Up to now, we only considered the *viscous damping*. With the viscous damping model, the damping force is proportional to the velocity (Fig. 1.1):

$$f_{\mathrm{v,d}} = -c\,\dot{x}(t). \tag{1.47}$$

However, not all physical processes can be assumed to behave according to this simplification. The equivalent viscous damping is used to relate the energy actually dissipated to the viscous damping equivalent. In case of viscous damping the work done by the damping force is:

$$W_{\mathrm{v,d}} = \int f_{\mathrm{v,d}}\,\mathrm{d}x = \int f_{\mathrm{v,d}}\,\dot{x}(t)\,\mathrm{d}t. \tag{1.48}$$

If harmonic response $x(t) = X\cos(\omega t)$ is assumed, then the work done per cycle is

$$\begin{aligned} W_{\mathrm{v,d,1}} &= \int_0^{2\pi/\omega} -c\,\dot{x}(t)^2\,\mathrm{d}t \\ &= \int_0^{2\pi/\omega} -c\,X^2\,\omega^2\,\sin^2(\omega t)\,\mathrm{d}t \\ &= -\pi\,c\,\omega\,X^2. \end{aligned} \tag{1.49}$$

If for a different damping model (DM), the work per cycle is defined with $W_{\mathrm{DM,1}}$ and this work should be equal to the equivalent viscous damping, then the *equivalent viscous damping* is defined as

$$c_{\mathrm{eq}} = \frac{W_{\mathrm{DM,1}}}{W_{\mathrm{v,d,1}}} = \frac{W_{\mathrm{DM,1}}}{-\pi\,\omega\,X^2}. \tag{1.50}$$

The viscous damping is proportional to the velocity (1.47); for harmonic oscillations at constant displacement amplitude, the velocity increases (and therefore the damping force/work) if the frequency of oscillation increases. Consequently, the viscous damping is said to be frequency dependent. However, in most real structures a displacement dependent damping force is observed (and not frequency). Due to this reason, the *structural* or also *hysteretic damping* d, which is proportional to displacement and in phase with velocity, was introduced. The structural damping force is defined as

$$f_{\mathrm{s,d}} = -\mathrm{i}\,d\,x(t), \tag{1.51}$$

where d is the parameter of structural/hysteretic damping, and the imaginary number i changes the phase.

If again harmonic response is assumed, then the work done per cycle is

$$\begin{aligned} W_{\mathrm{s,d,1}} &= \int_0^{2\pi/\omega} -\mathrm{i}\,d\,X\cos(\omega t)\,(-\omega\,X\sin(\omega t))\,\mathrm{d}t \\ &= \int_0^{2\pi/\omega} d\,X\sin(\omega t)\,(-\omega\,X\sin(\omega t))\,\mathrm{d}t \\ &= -\pi\,d\,X^2, \end{aligned} \tag{1.52}$$

where $\sin(\omega t)$ was used instead of $-\mathrm{i}\, d\, X \cos(\omega t)$ (in structural damping, i is used to achieve the phase delay). Finally, the viscous equivalent for structural damping is

$$\begin{aligned} c_{\mathrm{eq,s}} &= \frac{W_{\mathrm{s,d,1}}}{W_{\mathrm{v,d,1}}} \\ &= \frac{-\pi\, d\, X^2}{-\pi\, \omega\, X^2} \\ &= \frac{d}{\omega}. \end{aligned} \tag{1.53}$$

Using structural damping force (1.51) instead of the viscous (1.47) in the equation of motion (1.1) results in the nondimensional FRF (1.32):

$$\begin{aligned} H_0(\omega) &= \frac{k}{-\omega^2 m + \mathrm{i}\, d + k} \\ &= \frac{1}{1 - r^2 + \mathrm{i}\, \eta}, \end{aligned} \tag{1.54}$$

where

$$\eta = \frac{d}{k} \tag{1.55}$$

is the *damping loss factor*.

For structural damping, the maximum of $H_0 \cdot H_0^*$ (1.54) is at $r = 1$, whereas for viscous damping (1.32), the maximum is at $r = \sqrt{1 - 2\zeta^2}$. For lightly damped systems, we can assume that the maximum in both cases is at $r = 1$, and consequently from Eqs. (1.32) and (1.54) we have:

$$\begin{aligned} \frac{1}{1 - 1^2 + \mathrm{i}\, 2\, \zeta\, 1} &= \frac{1}{1 - 1^2 + \mathrm{i}\, \eta}, \\ \frac{1}{\mathrm{i}\, 2\, \zeta} &= \frac{1}{\mathrm{i}\, \eta}; \end{aligned} \tag{1.56}$$

at resonance, the relation between the damping ratio and damping loss factor is

$$\zeta = \frac{\eta}{2}. \tag{1.57}$$

1.1.3 Periodic excitation

Before the harmonic excitation discussed in Section 1.1.2 can be generalized to a general periodic excitation, the properties of the linear time-invariant system and the complex Fourier series need to be addressed.

Linear time-invariant system

The differential Eq. (1.1) can be considered as a system that relates the excitation (input) to the response (output):

$$\underbrace{f(t)}_{\text{Input}} \rightarrow \boxed{\text{System}} \rightarrow \underbrace{x(t)}_{\text{Output}} . \tag{1.58}$$

A system is considered *linear* if it has the properties of *additivity* and *scaling*[4]:

$$\begin{aligned} &\textit{additivity:} \quad f_1(t) + f_2(t) \quad \rightarrow \quad \boxed{\text{Linear system}} \quad \rightarrow \quad x_1(t) + x_2(t); \\ &\textit{scaling:} \qquad\qquad a\, f(t) \quad \rightarrow \quad \boxed{\text{Linear system}} \quad \rightarrow \quad a\, x(t). \end{aligned} \tag{1.59}$$

If both properties are valid, then we can apply the *superposition* principle:

$$a_1\, f_1(t) + a_2\, f_2(t) \quad \rightarrow \quad \boxed{\text{Linear system}} \quad \rightarrow \quad a_1\, x_1(t) + a_2\, x_2(t). \tag{1.60}$$

Another important property is the *time-invariance*:

$$f(t - t_0) \quad \rightarrow \quad \boxed{\text{Time-invariant system}} \quad \rightarrow \quad x(t - t_0). \tag{1.61}$$

Complex Fourier series

Fig. 1.6 shows a general periodic excitation force:

$$f(t) = f(t + T), \tag{1.62}$$

where T is the period. A real signal is shown; however, in signal processing, we frequently use complex functions for a general mathematical manipulation of real signals with arbitrary phase. Due to this, $f(t)$ is in general complex valued.

The periodic function $f(t)$ can be decomposed into harmonic components using the *complex Fourier series expansion*:

$$f(t) = \sum_{n=-\infty}^{+\infty} c_n\, \mathrm{e}^{\mathrm{i}\, n\, \omega_{\mathrm{p}}\, t}, \tag{1.63}$$

where ω_{p} is the fundamental angular frequency of the periodic function,

$$\omega_{\mathrm{p}} = \frac{2\pi}{T}. \tag{1.64}$$

4 Also called *homogeneity*.

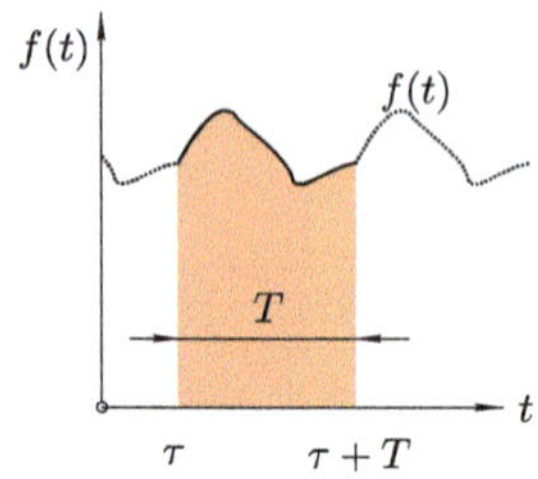

FIGURE 1.6 Periodic excitation force $f(t)$ with period T.

The Fourier (complex) coefficients c_n for $n \in \mathbb{Z}$ are defined by

$$c_n = \frac{1}{T} \int_{\tau}^{\tau+T} f(t)\, \mathrm{e}^{-\mathrm{i}\, n\, \omega_\mathrm{p}\, t}\, \mathrm{d}t, \tag{1.65}$$

where c_0 is the average value of $f(t)$. Further, for a real function $f(t)$, the Fourier coefficients are conjugate pairs:

$$c_{-n} = c_n^*, \tag{1.66}$$

where * is used to denote the complex conjugate.

Periodic excitation

Eq. (1.63) expanded the periodic excitation force to particular harmonics, and now the harmonic components can be solved individually and later superpositioned (1.60) to the final solution:

$$x(t) = \sum_{n=-\infty}^{+\infty} x_n(t), \tag{1.67}$$

where $x_n(t)$ is the steady-state time response of a particular harmonic n (1.21),

$$x_n(t) = X_n\, \mathrm{e}^{\mathrm{i}\, n\, \omega_\mathrm{p}\, t}, \tag{1.68}$$

where by (1.32)

$$\begin{aligned} X_n &= X_{0,n}\, H_0(\omega_\mathrm{p}\, n) \\ &= \frac{c_n}{k}\, H_0(\omega_\mathrm{p}\, n) \\ &= c_n\, \alpha(\omega_\mathrm{p}\, n), \end{aligned} \tag{1.69}$$

where $X_{0,n} = c_n/k$ is the "static deflection" term (1.31), and $\alpha(\omega) = H_0(\omega)/k$; see Eqs. (1.33) and (1.35).

1.1.4 General forcing excitation

Fourier integral

The complex Fourier series approach is based on the excitation period T; if this period is extended toward infinity, then the nonperiodic excitation can be evaluated. When the period approaches infinity, the Fourier complex series coefficient (1.65) transitions to the *Fourier integral.* Key steps in this transition are:

- The integration is set from $-T/2$ to $+T/2$:

$$c_n = \frac{1}{T}\int_{-T/2}^{T/2} f(t)\,\mathrm{e}^{-\mathrm{i}\,\omega_\mathrm{p}\,n\,t}\,\mathrm{d}t.$$

- The continuous angular frequency ω is used in obtaining the complex exponent c instead of $\omega_\mathrm{p}\,n$:

$$c(\omega) = \frac{1}{T}\int_{-T/2}^{T/2} f(t)\,\mathrm{e}^{-\mathrm{i}\,\omega t}\,\mathrm{d}t.$$

- Scale the integral with T:

$$c(\omega)\,T = \int_{-T/2}^{T/2} f(t)\,\mathrm{e}^{-\mathrm{i}\,\omega t}\,\mathrm{d}t.$$

- Let the period T approach infinity:

$$F(\omega) = \mathcal{F}\{f(t)\} = \int_{-\infty}^{\infty} f(t)\,\mathrm{e}^{-\mathrm{i}\,\omega t}\,\mathrm{d}t. \tag{1.70}$$

Eq. (1.70) presents the *forward Fourier transform* $\mathcal{F}$, and $F(\omega)$ is the complex amplitude at angular frequency ω.

Similarly, for the *inverse Fourier transform*, Eq. (1.63) is changed as the continuous angular frequency ω is used instead of $\omega_\mathrm{p}\,n$ and as T approaches infinity:

$$f(t) = \mathcal{F}^{-1}\{F(\omega)\} = \frac{1}{2\pi}\int_{-\infty}^{+\infty} F(\omega)\,\mathrm{e}^{\mathrm{i}\,\omega t}\,\mathrm{d}\omega. \tag{1.71}$$

The infinitesimal change of angular frequency is defined as $\mathrm{d}\omega = 2\pi\,\mathrm{d}t/T$. Eq. (1.71) presents the *inverse Fourier transform.* Eqs. (1.70) and (1.71) present the *Fourier transform pairs.*

Frequently, the Fourier transform is written in terms of the frequency $f = \omega/2\pi$ in Hz in a more symmetric form[5]:

$$F(f) = \mathcal{F}\{f(t)\} = \int_{-\infty}^{\infty} f(t)\, \mathrm{e}^{-\mathrm{i}\,2\pi\, f\, t}\, \mathrm{d}t, \tag{1.72}$$

$$f(t) = \mathcal{F}^{-1}\{F(f)\} = \int_{-\infty}^{\infty} F(f)\, \mathrm{e}^{\mathrm{i}\,2\pi\, f\, t}\, \mathrm{d}f, \tag{1.73}$$

where the symbols $\mathcal{F}$ and $\mathcal{F}^{-1}$ are used for the (forward) Fourier transform and the *inverse Fourier transform*.

With the Fourier transform, the superposition to periodic excitation can be generalized (see also Eqs. (1.67) and (1.69)):

$$x(t) = \mathcal{F}^{-1}\{X(\omega)\}, \tag{1.74}$$

where

$$X(\omega) = \alpha(\omega)\, F(\omega), \tag{1.75}$$

α is the FRF in the form of receptance (1.35), and $F(\omega)$ is the Fourier transform of the excitation force $\mathcal{F}\{f(t)\}$.

Unit-impulse force

For a general excitation shown in Fig. 1.7, the impulse force $f(\tau)$ is first researched using the theoretical *unit-impulse force excitation* shown in Fig. 1.8 and defined with the Dirac δ function:

$$f_1(t) = \delta(t - \tau), \tag{1.76}$$

which is zero for all values of t, except at τ, as the amplitude goes to infinity. The name *unit* comes from the integral of the force $f_1(t)$ over time t, which results in the *unit* change of the linear momentum ΔI:

$$\Delta I_1 = \lim_{\Delta\tau \to 0} \int_{\tau}^{\tau+\Delta\tau} f_1(t)\, \mathrm{d}t = 1. \tag{1.77}$$

At time τ^-, the system shown in Fig. 1.8 is at rest; at infinitesimal time later τ^+, due to change in unitary linear momentum ΔI_1, the velocity changes to

$$\dot{x}_0 = \frac{\Delta I_1}{m} = \frac{1}{m}. \tag{1.78}$$

[5] It is common to use f for frequency; however, here $f(t)$ is also used for the force in the time domain, and $F(f)$ is the force in the frequency domain.

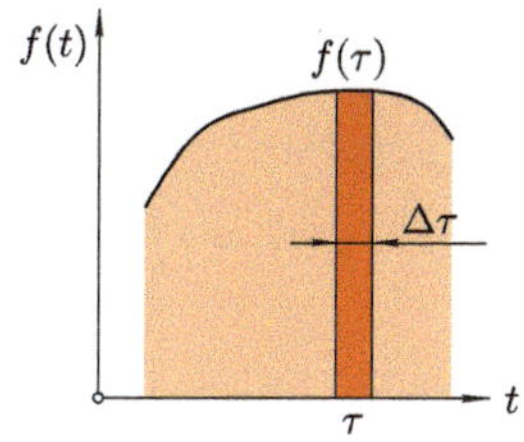

FIGURE 1.7 General force and the impulse force $f(\tau)$.

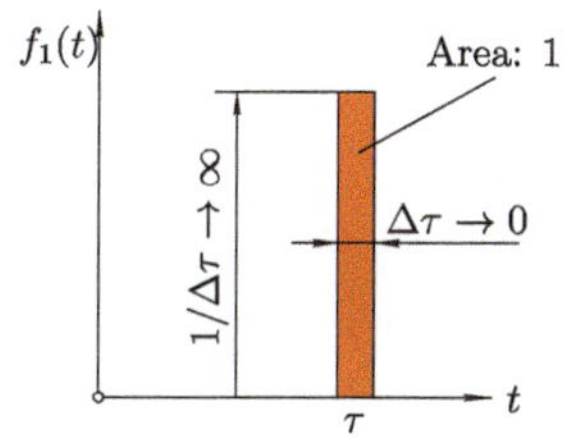

FIGURE 1.8 Unit-impulse force.

After the unit impulse $t \geq \tau$, the force $f_1(t)$ is zero, and the system responds with regard to the homogeneous response at zero initial displacement $x = 0$ and $\dot{x} = \dot{x}_0$; see solution defined by Eq. (1.17):

$$h(t-\tau) = x(t) = \frac{1}{m\,\omega_\mathrm{d}}\,\mathrm{e}^{-\zeta\,\omega_0\,(t-\tau)}\,\sin(\omega_\mathrm{d}\,(t-\tau)) \quad \text{for } t > \tau, \tag{1.79}$$

where $h(t-\tau)$ is the *unit impulse response function* (IRF).

The unit-impulse response of a *causal* system depends only on the past (and current) excitation, not the future. All systems in this book are considered causal: response is a result of past/current excitation.

The general force excitation shown in Fig. 1.7 can be seen as a superposition of impulses. Each infinitesimal impulse results in a response with regard to the unit impulse response (1.79) multiplied by the impulse amplitude $f(\tau)\,\Delta\tau$. For linear systems, the response can be superpositioned (1.60):

$$x(t) = \sum_{\tau} f(\tau)\,h(t-\tau)\,\Delta\tau \quad \text{for } t > \tau. \tag{1.80}$$

As $\Delta\tau \to 0$, the sum is replaced with an integral:

$$x(t) = \int_0^t f(\tau)\,h(t-\tau)\,\mathrm{d}\tau \quad \text{for } t > \tau. \tag{1.81}$$

Eq. (1.81) is called the *convolution/Duhamel's integral*.

As the IRF $h(t-\tau)=0$ for $t<\tau$, the lower limit in Eq. (1.81) can be extended to $-\infty$:

$$x(t)=\int_{-\infty}^{t} f(\tau)\,h(t-\tau)\,\mathrm{d}\tau \quad \text{for } t>\tau. \tag{1.82}$$

Further, by changing the variable of integration to $\tau_1=t-\tau$ ($\mathrm{d}\tau_1=-\mathrm{d}\tau$) the integration limits are: $-\infty\to+\infty$ and $t\to 0$:

$$x(t)=-\int_{+\infty}^{0} f(t-\tau_1)\,h(\tau_1)\,\mathrm{d}\tau_1=\int_{0}^{+\infty} f(t-\tau_1)\,h(\tau_1)\,\mathrm{d}\tau_1. \tag{1.83}$$

From Eqs. (1.82) and (1.83) the commutative property of the convolution can be summarized:

$$x(t)=\int_{-\infty}^{t} f(\tau)\,h(t-\tau)\,\mathrm{d}\tau=\int_{0}^{+\infty} h(\tau)\,f(t-\tau)\,\mathrm{d}\tau, \tag{1.84}$$

which is written in short form as

$$x(t)=f(t)*h(t)=h(t)*f(t), \tag{1.85}$$

where $*$ denotes the *convolution*.

Note that if the system is *noncausal* and the response depends on future input, then the integration limits are extended, and the commutative property of the convolution is

$$x(t)=\int_{-\infty}^{+\infty} f(t-\tau)\,h(\tau)\,\mathrm{d}\tau=\int_{-\infty}^{+\infty} f(\tau)\,h(t-\tau)\,\mathrm{d}\tau. \tag{1.86}$$

Frequency response vs unit-impulse response

In the previous section the unit-impulse response $h(t)$ and the complex frequency response $\alpha(\omega)$ were introduced. Whereas the first describes the linear system in the time domain, the later describes it in the frequency domain. We will now show that $h(t)$ and $\alpha(\omega)$ form the Fourier transform pair.

The Fourier transform of the unit-impulse force $f_1(t)=\delta(t)$ (1.76) is

$$F_1(\omega)=\mathcal{F}\{f_1(t)\}=\int_{-\infty}^{\infty} f_1(t)\,\mathrm{e}^{-\mathrm{i}\,\omega t}\,\mathrm{d}t=1, \tag{1.87}$$

and consequently the unit-impulse excites all frequencies equally.

By Eq. (1.75) the response in the frequency domain to a unit excitation is

$$X_1(\omega)=\alpha(\omega)\,\underbrace{F_1(\omega)}_{=1}=\alpha(\omega). \tag{1.88}$$

Applying the inverse Fourier transform to the response, we have

$$\begin{aligned} x_1(t) &= \mathcal{F}^{-1}\{X_1(\omega)\} \\ &= \mathcal{F}^{-1}\{\alpha(\omega)\}. \end{aligned} \tag{1.89}$$

On the other hand, the response is defined with the unit-impulse response function (1.79): $x_1(t) = h(t)$. Finally, it follows that

$$h(t) = \mathcal{F}^{-1}\{\alpha(\omega)\} \qquad \text{or} \qquad \alpha(\omega) = \mathcal{F}\{h(t)\}. \tag{1.90}$$

This confirms that the unit-impulse response and the complex frequency response form a Fourier pair.

Pole residual Fourier pair

The inverse Fourier transform of the receptance in the partial expansion form (1.45) is

$$h(t) = \mathcal{F}^{-1}\{\alpha(\omega)\} = R\,e^{\lambda t} + R^*\,e^{\lambda^* t}, \tag{1.91}$$

where R is the pole residue (1.46), and λ is the complex eigenvalue (1.13).

1.1.5 Base excitation

Fig. 1.9 shows a system with two masses, m and m_b, where the latter is considered the base. When *base motion/excitation* is considered, then the kinematics of the base $y(t)$ is defined in time and therefore does not present a degree of freedom; the only degree of freedom is the response at $x(t)$. This is typical when a relatively small mass is attached to a relatively big mass (e.g., a starter attached to an engine). In this case the assumption is made that the dynamics of the base is not changed due to the dynamics of the response mass; consequentially, the kinematics measured at the base can be considered the input to the system. Base excitation is usually used at accelerated vibration testing with electrodynamic shakers. The vibration controller (with a feedback loop) ensures the required kinematics of the base.

On the other hand, if the excitation force $f(t)$ is defined, then the system has two degrees of freedom, $x(t)$ and $y(t)$. This corresponds to the accelerated vibration testing without a feedback loop controller (e.g., constant power spectral density voltage excitation).

If the kinematics $y(t)$ is defined, then this is also called *kinematic excitation*, whereas if the force $f(t)$ is defined, then this is called *force* or *dynamic excitation*. From the vibration fatigue point of view, the type of excitation is very important as the natural frequencies are different if the system is kinematically or dynamically excited. Here we will discuss the kinematically excited, single-degree of freedom, system, whereas in Section 1.2.4, we will discuss the dynamically excited, two-degree of freedom, system.

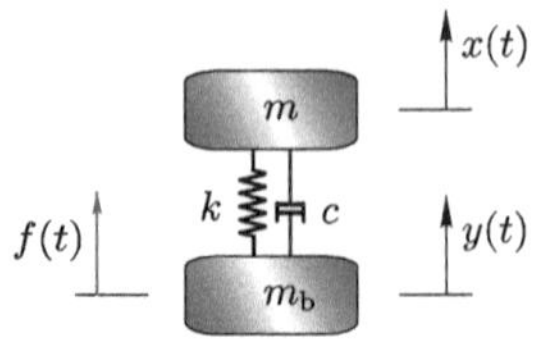

FIGURE 1.9 Excitation of a two-mass system.

The equations of motion in case of kinematic excitation can be obtained from the free-body diagram (see Fig. 1.10):

$$-k\,(x(t)-y(t))-c\,(\dot{x}(t)-\dot{y}(t))=m\,\ddot{x}(t) \tag{1.92}$$

and, after separation of variables,

$$m\,\ddot{x}(t)+c\,\dot{x}(t)+k\,x(t)=c\,\dot{y}(t)+k\,y(t). \tag{1.93}$$

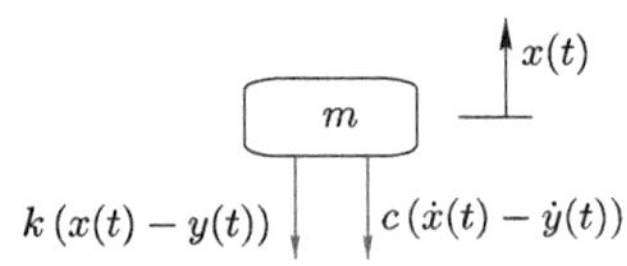

FIGURE 1.10 A single-degree-of-freedom model: free-body diagram.

Similarly, as before, we assume harmonic excitation and response:

$$y(t)=Y\,\mathrm{e}^{\mathrm{i}\,\omega\,t}, \qquad x(t)=X\,\mathrm{e}^{\mathrm{i}\,\omega\,t}, \tag{1.94}$$

where X and Y are complex amplitudes. With the harmonic assumption, Eq. (1.93) changes to

$$-\omega^2\,m\,X\,\mathrm{e}^{\mathrm{i}\,\omega\,t}+\mathrm{i}\,\omega\,c\,X\,\mathrm{e}^{\mathrm{i}\,\omega\,t}+k\,X\,\mathrm{e}^{\mathrm{i}\,\omega\,t}=\mathrm{i}\,\omega\,c\,Y\,\mathrm{e}^{\mathrm{i}\,\omega\,t}+k\,Y\,\mathrm{e}^{\mathrm{i}\,\omega\,t}. \tag{1.95}$$

Since $\mathrm{e}^{\mathrm{i}\,\omega\,t}\neq 0$, Eq. (1.95) simplifies to

$$-\omega^2\,m\,X+\mathrm{i}\,\omega\,c\,X+k\,X=\mathrm{i}\,\omega\,c\,Y+k\,Y. \tag{1.96}$$

At base excitation the complex frequency response function is

$$\frac{X}{Y}=\frac{\mathrm{i}\,\omega\,c+k}{-\omega^2\,m+\mathrm{i}\,\omega\,c+k}=\frac{1+\mathrm{i}\,2\,\zeta\,r}{1-r^2+\mathrm{i}\,2\,\zeta\,r}. \tag{1.97}$$

The frequency response amplitude and phase at kinematic excitation are shown in Fig. 1.11.

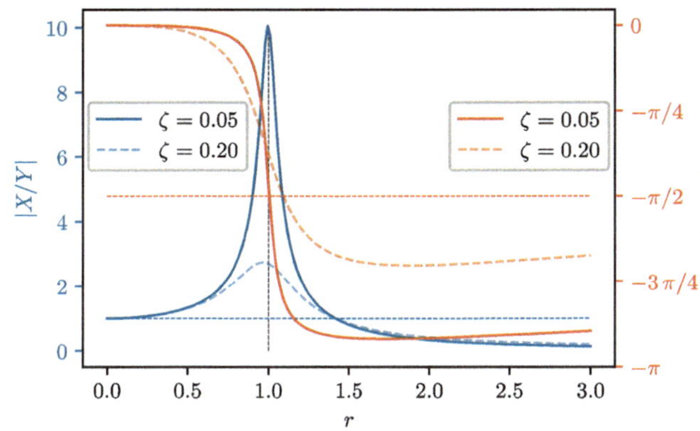

FIGURE 1.11 Amplitude and phase of the frequency response function of an SDOF model for base excitation.

However, if the relative motion is considered $z(t) = x(t) - y(t)$, then the equation of motion (1.92) changes to

$$-k\,z(t) - c\,\dot{z}(t) = m\,\ddot{z}(t) + m\,\ddot{y}(t). \tag{1.98}$$

After harmonic assumption, the nontrivial solution is

$$-\omega^2 m\,Z + \mathrm{i}\,\omega\,c\,Z + k\,Z = \omega^2 m\,Y. \tag{1.99}$$

Therefore the relative frequency response function is

$$\frac{Z}{Y} = \frac{\omega^2 m}{-\omega^2 m + \mathrm{i}\,\omega\,c + k} = \frac{r^2}{1 - r^2 + \mathrm{i}\,2\,\zeta\,r}. \tag{1.100}$$

The relative frequency response amplitude and phase at kinematic excitation are shown in Fig. 1.12.

The significance of base excitation with regard to vibration fatigue is shown in Fig. 1.12: the relative complex amplitude Z significantly increases close to the natural frequency ($r = 1$), and $|Z/Y|$ converges to 1 above the natural frequency. It is worth noting that at kinematic excitation the undamped natural frequency is defined with $\omega_0 = \sqrt{k/m}$ (1.9).

1.2 Multi-degree-of-freedom systems

In vibration fatigue, we are usually interested in the *continuous systems* that have infinite number of degrees-of-freedom. The numerical models of those continuous systems have however finite number of degrees-of-freedom and are in general considered *multi-degree-of-freedom systems* (MDOF). This section covers the basic MDOF systems as a generalization of SDOF systems (see Section 1.1).

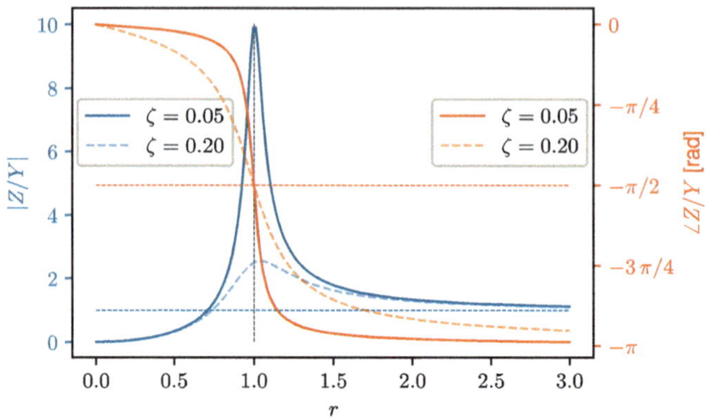

FIGURE 1.12 Amplitude and phase of the frequency response function of an SDOF model for base excitation (relative motion).

Fig. 1.13 shows a general N degrees of freedom system described in a general form of the equations of motion as

$$\mathbf{M}\,\ddot{\mathbf{x}}(t) + \mathbf{C}\,\dot{\mathbf{x}}(t) + \mathbf{K}\,\mathbf{x}(t) = \mathbf{f}(t), \tag{1.101}$$

where **M**, **C**, and **K** are the *mass*, *viscous damping*, and *stiffness* matrices of dimension $N \times N$, respectively, $\mathbf{f}(t)$ is the *force load vector*, and $\mathbf{x}(t)$ is the *displacement vector*:

$$\mathbf{x}(t) = \left[x_1(t), x_2(t), \ldots, x_N(t)\right]^{\mathrm{T}}. \tag{1.102}$$

For details on different approaches on how to derive the equations of motion, we refer the interested reader to [1–3].

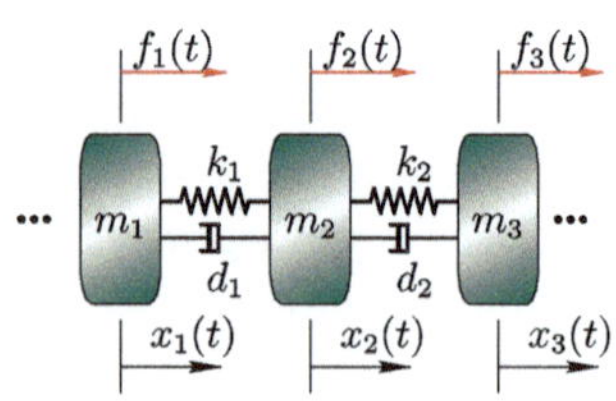

FIGURE 1.13 A multi-degree-of-freedom system model.

1.2.1 Undamped free response

The solution to Eq. (1.101) is first researched for the undamped ($\mathbf{C} = \mathbf{0}$) free response ($\mathbf{f}(t) = \mathbf{0}$):

$$\mathbf{M}\,\ddot{\mathbf{x}}(t) + \mathbf{K}\,\mathbf{x}(t) = \mathbf{0}. \tag{1.103}$$

Similarly to the SDOF case (1.4), we assume a harmonic response

$$\mathbf{x}(t) = \mathbf{X}\,\mathrm{e}^{\lambda t}, \tag{1.104}$$

where $\mathbf{X}$ is the vector of displacement amplitudes (in general complex). With the harmonic assumption, Eq. (1.103) results in

$$\left(\lambda^2 \mathbf{M} + \mathbf{K}\right)\mathbf{X}\,\mathrm{e}^{\lambda t} = \mathbf{0}. \tag{1.105}$$

As $\exp(\lambda t) \neq 0$ for all λ and t, this simplifies to the *eigenvalue problem*

$$\left(\mathbf{K} + \lambda^2 \mathbf{M}\right)\mathbf{X} = \mathbf{0}. \tag{1.106}$$

A nontrivial solution $\mathbf{X} \neq \mathbf{0}$ of the eigenvalue problem can be obtained with the characteristic equation

$$\det\left(\mathbf{K} + \lambda^2 \mathbf{M}\right) = 0. \tag{1.107}$$

For an N-DOF system, we obtain N eigenvalues λ_r^2. For the undamped case, $\lambda_r^2 = \omega_r^2$, where ω_r is the (angular) natural frequency.[6] Further, for each natural frequency ω_r, we can obtain the corresponding *eigenvector/natural mode/mode shape*

$$\boldsymbol{\Psi}_r = \begin{pmatrix} \Psi_{r,1} \\ \Psi_{r,2} \\ \vdots \\ \Psi_{r,N} \end{pmatrix}, \tag{1.108}$$

where $\Psi_{r,i}$ is the amplitude at the rth natural frequency at the ith degree of freedom. The elements $\Psi_{r,i}$ are not uniquely determined. If, for example, the first component is set to $\Phi_{r,1} = 1$, then all other components $\Phi_{r,i}$ are uniquely defined.[7]

Another approach to obtain the eigenvalue problem of the MDOF system is multiplying Eq. (1.106) by $\mathbf{M}^{-1}$, which results in

$$\left(\mathbf{M}^{-1}\mathbf{K} + \lambda^2 \mathbf{I}\right)\mathbf{X} = \mathbf{0}. \tag{1.109}$$

Alternatively, the stiffness matrix $\mathbf{K}$ is inverted, and the eigenproblem is

$$\left(\mathbf{K}^{-1}\mathbf{M} + \frac{1}{\lambda^2}\mathbf{I}\right)\mathbf{X} = \mathbf{0}. \tag{1.110}$$

6 Eigenvalues are sometimes called *squared natural frequencies*.

7 This opens up the possibilities for different types of scaling/normalizing, which will be discussed later in this section.

The term $\mathbf{K}^{-1}\mathbf{M}$ is also called the *dynamic matrix* and sometimes denoted by $\mathbf{D}$. Solving an eigensystem is a typical numerical problem, and for *Hermitian* (conjugate symmetric) and *sparse* matrices, significant speed ups are possible [5].

Orthogonality of eigenvectors

For the particular natural frequency ω_i and the corresponding mode shape vector $\mathbf{\Psi}_i$, Eq. (1.106) can be rewritten as

$$\mathbf{K}\,\mathbf{\Psi}_i = \omega_i^2\,\mathbf{M}\,\mathbf{\Psi}_i. \tag{1.111}$$

Multiplying (1.111) with $\mathbf{\Psi}_j^{\mathrm{T}}$ from the left-hand side, we have

$$\mathbf{\Psi}_j^{\mathrm{T}}\,\mathbf{K}\,\mathbf{\Psi}_i = \omega_i^2\,\mathbf{\Psi}_j^{\mathrm{T}}\,\mathbf{M}\,\mathbf{\Psi}_i. \tag{1.112}$$

The mass $\mathbf{M}$ and stiffness $\mathbf{K}$ matrices are symmetric, and transposing Eq. (1.112) results in

$$\mathbf{\Psi}_i^{\mathrm{T}}\,\mathbf{K}\,\mathbf{\Psi}_j = \omega_i^2\,\mathbf{\Psi}_i^{\mathrm{T}}\,\mathbf{M}\,\mathbf{\Psi}_j. \tag{1.113}$$

Similarly, for the natural frequency ω_j and the corresponding mode shape vector $\mathbf{\Psi}_j$, Eq. (1.106) results in

$$\mathbf{K}\,\mathbf{\Psi}_j = \omega_j^2\,\mathbf{M}\,\mathbf{\Psi}_j. \tag{1.114}$$

If Eq. (1.114) is multiplied from the left-hand side by $\mathbf{\Psi}_i^{\mathrm{T}}$, then

$$\mathbf{\Psi}_i^{\mathrm{T}}\,\mathbf{K}\,\mathbf{\Psi}_j = \omega_j^2\,\mathbf{\Psi}_i^{\mathrm{T}}\,\mathbf{M}\,\mathbf{\Psi}_j. \tag{1.115}$$

If Eq. (1.113) is subtracted from Eq. (1.115), then

$$0 = (\omega_j^2 - \omega_i^2)\,\left(\mathbf{\Psi}_i^{\mathrm{T}}\,\mathbf{M}\,\mathbf{\Psi}_j\right). \tag{1.116}$$

In general, $(\omega_j^2 - \omega_i^2) \neq 0$, and therefore[8]

$$0 = \mathbf{\Psi}_i^{\mathrm{T}}\,\mathbf{M}\,\mathbf{\Psi}_j \qquad \text{for} \quad i \neq j, \tag{1.117}$$

which confirms that the ith and jth mode shapes are *orthogonal* with respect to *mass*.

Inserting Eq. (1.113) into (1.117), we have

$$0 = \mathbf{\Psi}_i^{\mathrm{T}}\,\mathbf{K}\,\mathbf{\Psi}_j \qquad \text{for} \quad i \neq j, \tag{1.118}$$

which confirms that the ith and jth mode shapes are orthogonal with respect to *stiffness*.

[8] Special care is required in case of closely spaced frequencies; see, e.g., [1].

Due to the orthogonality with respect to mass and stiffness, Eq. (1.112) can be rewritten for the case $r = i = j$:

$$\omega_r^2 = \frac{\mathbf{\Psi}_r^{\mathrm{T}}\,\mathbf{K}\,\mathbf{\Psi}_r}{\mathbf{\Psi}_r^{\mathrm{T}}\,\mathbf{M}\,\mathbf{\Psi}_r} = \frac{k_r}{m_r}, \tag{1.119}$$

where k_r and m_r are the *modal* or *generalized stiffness* and *mass* of the mode r, respectively:

$$\begin{aligned} k_r &= \mathbf{\Psi}_r^{\mathrm{T}}\,\mathbf{K}\,\mathbf{\Psi}_r, \\ m_r &= \mathbf{\Psi}_r^{\mathrm{T}}\,\mathbf{M}\,\mathbf{\Psi}_r. \end{aligned} \tag{1.120}$$

Normalization of modes

The eigenvector $\mathbf{\Psi}_r$ (1.108) is uniquely defined to an arbitrary *scaling factor* γ_i:

$$\mathbf{\Phi}_r = \gamma_r\,\mathbf{\Psi}_r, \tag{1.121}$$

where $\mathbf{\Psi}_r$ is considered to be arbitrary, and $\mathbf{\Phi}_r$ is considered to be scaled. Inserting (1.121) into (1.119), we have

$$\omega_r^2 = \frac{k_r}{m_r} = \frac{\gamma_r^2\,\mathbf{\Phi}_r^{\mathrm{T}}\,\mathbf{K}\,\mathbf{\Phi}_r}{\gamma_r^2\,\mathbf{\Phi}_r^{\mathrm{T}}\,\mathbf{M}\,\mathbf{\Phi}_r} = \frac{\mathbf{\Phi}_r^{\mathrm{T}}\,\mathbf{K}\,\mathbf{\Phi}_r}{\mathbf{\Phi}_r^{\mathrm{T}}\,\mathbf{M}\,\mathbf{\Phi}_r}, \tag{1.122}$$

which shows that the scaling factor does not change the natural frequency, but can be used for different types of eigenvector normalizations. In structural dynamics the *mass normalization* is typically used[9]:

$$m_r = \mathbf{\Phi}_r^{\mathrm{T}}\,\mathbf{M}\,\mathbf{\Phi}_r = 1. \tag{1.123}$$

By (1.123) and (1.120) the scaling factor is

$$\gamma_r = \frac{1}{\sqrt{\mathbf{\Psi}_r^{\mathrm{T}}\,\mathbf{M}\,\mathbf{\Psi}_r}}. \tag{1.124}$$

With mass normalization $m_r = 1$, by (1.122) the stiffness is defined as

$$k_r = \omega_r^2. \tag{1.125}$$

An important consequence of the mass normalization is the generalization of the receptance of the rth mode, where the SDOF equation (1.28) simplifies to

$$\alpha_r(\omega) = \frac{1}{\omega_r^2 - \omega^2 + \mathrm{i}\,2\,\zeta_r\,\omega\,\omega_r} \tag{1.126}$$

[9] In this book, we only use mass normalization. $\mathbf{\Phi}$ will always refer to mass normalized eigenvectors/mode shapes. *Stiffness normalization*, which normalizes the eigenvectors with regard to $k_r = 1$, is not used in this book.

for viscous damping and to

$$\alpha_r(\omega) = \frac{1}{\omega_r^2 - \omega^2 + \mathrm{i}\,\eta_r\,\omega_r^2} \tag{1.127}$$

for the hysteretic damping.

Modal matrix and modal expansion

With the introduction of (mass normalized) *modal matrix*

$$\mathbf{\Phi} = \left[\mathbf{\Phi}_1, \mathbf{\Phi}_2, \cdots, \mathbf{\Phi}_i, \cdots, \mathbf{\Phi}_N\right], \tag{1.128}$$

the response in *physical coordinates* $\mathbf{x}(t)$ can be defined as

$$\mathbf{x}(t) = \mathbf{\Phi}\,\mathbf{q}(t), \tag{1.129}$$

where

$$\mathbf{q}(t) = [q_1(t), q_2(t), \ldots, q_i(t) \ldots, q_N(t)]^{\mathrm{T}} \tag{1.130}$$

are the *modal* (or *principal*) coordinates. Using the modal coordinates in the equation of motion (1.103) and multiplying them from left by $\mathbf{\Phi}^{\mathrm{T}}$, we get

$$\mathbf{\Phi}^{\mathrm{T}}\,\mathbf{M}\,\mathbf{\Phi}\,\ddot{\mathbf{q}}(t) + \mathbf{\Phi}^{\mathrm{T}}\,\mathbf{K}\,\mathbf{\Phi}\,\mathbf{q}(t) = \mathbf{0}. \tag{1.131}$$

Due to the orthogonality (1.117–(1.118) and mass normalization (1.123) ($m_r = 1$, $k_r = \omega_r^2$), the equation of motion simplifies to

$$\begin{bmatrix} \ddots & & \\ & 1 & \\ & & \ddots \end{bmatrix} \ddot{\mathbf{q}}(t) + \begin{bmatrix} \ddots & & \\ & \omega_r^2 & \\ & & \ddots \end{bmatrix} \mathbf{q}(t) = \mathbf{0}, \tag{1.132}$$

where the modal mass and stiffness matrices become diagonal (1.120), and the equations of motion become uncoupled. Consequently, each differential equation can be solved separately:

$$\ddot{q}_r(t) + \omega_r^2\,q_r(t) = 0. \tag{1.133}$$

Similarly to the homogeneous solution of an SDOF system (1.15), the solution for $q_r(t)$ can be written and depends on the initial conditions.

1.2.2 Damped free response

In vibration fatigue, we are in general not interested in the free response, but rather in the forced response. Therefore here we only give the necessary theoretical background required later in the forced response. We refer the interested reader to [1,2,4] for details.

Viscously damped system

After the free response of an undamped system has been discussed, we consider the free response of the damped system:

$$\mathbf{M}\,\ddot{\mathbf{x}}(t) + \mathbf{C}\,\dot{\mathbf{x}}(t) + \mathbf{K}\,\mathbf{x}(t) = \mathbf{0}. \tag{1.134}$$

Similarly to (1.131), using the modal coordinates in the equation of motion (1.134) and multiplying from left by $\boldsymbol{\Phi}^{\mathrm{T}}$, we have

$$\boldsymbol{\Phi}^{\mathrm{T}}\,\mathbf{M}\,\boldsymbol{\Phi}\,\ddot{\mathbf{q}}(t) + \boldsymbol{\Phi}^{\mathrm{T}}\,\mathbf{C}\,\boldsymbol{\Phi}\,\dot{\mathbf{q}}(t) + \boldsymbol{\Phi}^{\mathrm{T}}\,\mathbf{K}\,\boldsymbol{\Phi}\,\mathbf{q}(t) = \mathbf{0}. \tag{1.135}$$

In general, $\boldsymbol{\Phi}^{\mathrm{T}}\,\mathbf{C}\,\boldsymbol{\Phi}$ is not diagonal, and therefore the equations of motion stay coupled:

$$\begin{bmatrix} \ddots & & \\ & 1 & \\ & & \ddots \end{bmatrix} \ddot{\mathbf{q}}(t) + \boldsymbol{\Phi}^{\mathrm{T}}\,\mathbf{C}\,\boldsymbol{\Phi}\,\dot{\mathbf{q}}(t) + \begin{bmatrix} \ddots & & \\ & \omega_r^2 & \\ & & \ddots \end{bmatrix} \mathbf{q}(t) = \mathbf{0}. \tag{1.136}$$

Proportional viscous damping

For lightly damped systems, we introduce the *proportional* (or *Rayleigh*) damping to uncouple the equations of motion. We achieve the uncoupling by introducing a damping matrix proportional to the mass and/or stiffness matrix:

$$\mathbf{C} = \alpha\,\mathbf{M} + \beta\,\mathbf{K}, \tag{1.137}$$

where α and β are the mass and stiffness factors, respectively. Consequently, Eq. (1.136) simplifies to

$$\begin{bmatrix} \ddots & & \\ & 1 & \\ & & \ddots \end{bmatrix} \ddot{\mathbf{q}}(t) + \begin{bmatrix} \ddots & & \\ & \alpha + \beta\,\omega_r^2 & \\ & & \ddots \end{bmatrix} \dot{\mathbf{q}}(t) + \begin{bmatrix} \ddots & & \\ & \omega_r^2 & \\ & & \ddots \end{bmatrix} \mathbf{q}(t) = \mathbf{0}. \tag{1.138}$$

As the equations are now uncoupled, we have

$$\ddot{q}_r(t) + (\alpha + \beta\,\omega_r^2)\,\dot{q}_r(t) + \omega_r^2\,q_r(t) = 0. \tag{1.139}$$

If these uncoupled differential equations are compared to the SDOF differential equation (1.12),

$$\alpha + \beta\,\omega_r^2 = 2\,\zeta_r\,\omega_r, \tag{1.140}$$

then the mass and stiffness parameters of the proportional damping can be related to the damping ratio of a particular mode r with the undamped natural frequency ω_r:

$$\zeta_r = \frac{\alpha}{2\,\omega_r} + \frac{\beta\,\omega_r}{2}. \tag{1.141}$$

General viscously damped system

In general, the damping is not proportional, and therefore the damping matrix **C** in Eq. (1.134) is nondiagonal. Assume that a general harmonic solution is

$$\mathbf{x}(t) = \mathbf{X}\,\mathrm{e}^{\lambda\,t}, \tag{1.142}$$

where **X** is a vector of complex amplitudes, and λ is a complex number. Inserting (1.142) into (1.134) leads to a complex eigenproblem

$$\left(\lambda^2\,\mathbf{M} + \lambda\,\mathbf{C} + \mathbf{K}\right)\,\mathbf{X} = \mathbf{0}, \tag{1.143}$$

which is solved using the *state vector*

$$\mathbf{u}(t) = \begin{bmatrix} \mathbf{x}(t) \\ \dot{\mathbf{x}}(t) \end{bmatrix}. \tag{1.144}$$

Eq. (1.134) can now be rewritten as

$$\begin{bmatrix} \mathbf{C} & \mathbf{M} \\ \mathbf{M} & \mathbf{0} \end{bmatrix} \dot{\mathbf{u}}(t) + \begin{bmatrix} \mathbf{K} & \mathbf{0} \\ \mathbf{0} & -\mathbf{M} \end{bmatrix} \mathbf{u}(t) = \mathbf{0} \tag{1.145}$$

or written with matrices **A** and **B** (of shape $2N \times 2N$) as

$$\mathbf{A}\,\dot{\mathbf{u}}(t) + \mathbf{B}\,\mathbf{u}(t) = \mathbf{0}. \tag{1.146}$$

Combining Eqs. (1.142) and (1.144), we can rewrite the state vector as

$$\mathbf{u}(t) = \begin{bmatrix} \mathbf{X} \\ \lambda\,\mathbf{X} \end{bmatrix} \mathrm{e}^{\lambda\,t} = \mathbf{U}\,\mathrm{e}^{\lambda\,t}, \tag{1.147}$$

where **U** is the vector of state space complex amplitudes. As $\dot{\mathbf{u}}(t) = \lambda\,\mathbf{u}(t)$, the equations of motion (1.146) are rewritten as

$$\left[\lambda\mathbf{A} + \mathbf{B}\right]\,\mathbf{U} = \mathbf{0}, \tag{1.148}$$

which represents the generalized eigenproblem; in case of underdamped system the $2\,N$ eigenvalues are complex conjugate. If N eigenvalues are denoted as λ_r and the remaining N conjugated pairs as λ_r^*, then the corresponding eigenvectors are $\tilde{\mathbf{\Psi}}_r$ and $\tilde{\mathbf{\Psi}}_r^*$, respectively:

$$\tilde{\mathbf{\Psi}}_r = \begin{bmatrix} \mathbf{\Psi}_r \\ \lambda_r\,\mathbf{\Psi}_r \end{bmatrix}, \qquad \tilde{\mathbf{\Psi}}_r^* = \begin{bmatrix} \mathbf{\Psi}_r^* \\ \lambda_r^*\,\mathbf{\Psi}_r^* \end{bmatrix}, \tag{1.149}$$

where $\mathbf{\Psi}_r$ and its complex conjugate $\mathbf{\Psi}_r^*$ are the $N \times 1$ complex eigenvectors corresponding to the vector space coordinates $\mathbf{x}(t)$ (1.142).

Similarly to the undamped (see (1.117) and (1.118)) or proportionally damped system, the orthogonality of eigenvectors can be proven for general viscous damping. Consequently, the $2N$ equations of motion become uncoupled. As the mathematical deduction is quite lengthy, here we only give a summary (see, e.g., [4] for details). Using the analogy of an SDOF system, Eqs. (1.13) and (1.14), the eigenvalue can be written as (see also (1.13))

$$\lambda_r = -\omega_r\,\zeta_r + \mathrm{i}\,\omega_r\,\sqrt{1-\zeta_r^2}, \tag{1.150}$$

where by the orthogonality of eigenvectors the natural frequencies ω_r and damping ratios ζ_r are defined by

$$\frac{\boldsymbol{\Psi}_r^{*\mathrm{T}}\,\mathbf{K}\,\boldsymbol{\Psi}_r}{\boldsymbol{\Psi}_r^{*\mathrm{T}}\,\mathbf{M}\,\boldsymbol{\Psi}_r} = \frac{k_r}{m_r} = \omega_r^2 \tag{1.151}$$

and

$$\frac{\boldsymbol{\Psi}_r^{*\mathrm{T}}\,\mathbf{C}\,\boldsymbol{\Psi}_r}{\boldsymbol{\Psi}_r^{*\mathrm{T}}\,\mathbf{M}\,\boldsymbol{\Psi}_r} = \frac{c_r}{m_r} = 2\,\omega_r\,\zeta_r. \tag{1.152}$$

It is worth noting here that the eigenvectors $\boldsymbol{\Psi}$ are complex valued, and consequentially the degrees of freedom are in general not in phase (which is the case in undamped (1.104) or proportionally damped MDOF system).

Structural or hysteretic damping

An MDOF system with hysteretic damping is defined as

$$\mathbf{M}\,\ddot{\mathbf{x}}(t) + \mathrm{i}\,\mathbf{D}\,\mathbf{x}(t) + \mathbf{K}\,\mathbf{x}(t) = \mathbf{0}, \tag{1.153}$$

where the hysteretic damping matrix $\mathbf{D}$ is of dimension $N \times N$.

Proportional hysteretic damping

In general, the hysteretic damping results in coupled equations of motion. Similarly as in viscous damping, we introduce proportional hysteretic damping:

$$\mathbf{D} = \epsilon\,\mathbf{K} + \nu\,\mathbf{M}, \tag{1.154}$$

where ϵ and ν are the stiffness and mass factors, respectively. Using Eqs. (1.142) and (1.154) with Eq. (1.153),

$$\left(\mathbf{K} + \lambda^2\,\mathbf{M}\right) + \mathrm{i}\,\left(\epsilon\,\mathbf{K} + \nu\,\mathbf{M}\right) = \mathbf{0} \tag{1.155}$$

results in a complex eigenvalue problem, which in turn results in an uncoupled system of differential equations [4], where the eigenvectors are the same as for the undamped case (1.13). The complex eigenvalues are defined as

$$\lambda_r^2 = -\omega_r^2\,(1 + \mathrm{i}\,\eta_r)\,, \tag{1.156}$$

where the natural frequency ω_r is defined by

$$\omega_r^2 = \frac{k_r}{m_r}. \tag{1.157}$$

The modal stiffness k_r and mass m_r are defined by Eqs. (1.120). In Eq. (1.156) the *damping loss factor* is defined as

$$\eta_r = \epsilon + \frac{\nu}{\omega_r^2}. \tag{1.158}$$

General hysteretic damped system

If the hysteretic damping is not proportional, then the eigenvalues λ_r and eigenvectors $\mathbf{X}_r$ are complex valued. The eigenvalue is defined as [4]

$$\lambda_r^2 = \frac{\boldsymbol{\Psi}_r^{*\mathrm{T}}\,(\mathbf{K} + \mathrm{i}\,\mathbf{D})\,\boldsymbol{\Psi}_r}{\boldsymbol{\Psi}_r^{*\mathrm{T}}\,\mathbf{M}\,\boldsymbol{\Psi}_r} = \frac{k_r}{m_r}, \tag{1.159}$$

where k_r and m_r are the complex modal stiffness and mass, respectively.

Modal damping, free response

Frequently, instead of the viscous or hysteretic model, the modal damping approach is used. This approach assumes that the MDOF system (1.136) damping matrix $\boldsymbol{\Phi}^{\mathrm{T}}\,\mathbf{C}\,\boldsymbol{\Phi}$ can be diagonalized. By using the mass normalization we can deduce the following mode shape based equation:

$$\ddot{q}_r(t) + 2\,\zeta_r\,\omega_r\,\dot{q}_r(t) + \omega_r^2\,q_r(t) = 0, \tag{1.160}$$

where the mass normalized and diagonalized damping elements have been denoted by $2\,\zeta_r\,\omega_r$ (similarly to the SDOF case; see Eq. (1.12)); with this, each modal component requires a damping ratio ζ_r, which can be related to different damping models as shown in Table 1.2.

Similarly to the homogeneous solution of an SDOF system (1.15), we have

$$q_r(t) = Q_r\,\mathrm{e}^{\lambda_r\,t} + Q_{r+N}\,\mathrm{e}^{\lambda_r^*\,t}, \tag{1.161}$$

where the $2N$ unknown *scaling factors* (or *modal participation factors*) Q_r are defined from the $2N$ initial conditions. To apply the initial conditions, the modal coordinates need to be related to the physical (1.129):

$$\mathbf{x}(t) = \sum_{r=1}^{N} Q_r\,\Psi_r\,\mathrm{e}^{\lambda_r\,t} + Q_{r+N}\,\Psi_r^*\,\mathrm{e}^{\lambda_r^*\,t}, \tag{1.162}$$

where Ψ_r is the rth mode shape.[10]

---

[10] Instead of Ψ_r, the mass normalized mode shapes Φ_r could be used (resulting in a different scaling factor Q_r).

TABLE 1.2 Damping ratio vs. the damping model.

Damping model	Damping ratio ζ_r
Proportional viscous damping (1.141)	$\frac{\alpha}{2\,\omega_r} + \frac{\beta\,\omega_r}{2}$
General viscous damping (1.151)–(1.152)	$\frac{c_r}{2\,\omega_r\,m_r}$
Proportional hysteretic (1.57)–(1.158)	$\frac{1}{2}\left(\epsilon + \frac{\nu}{\omega_r^2}\right)$

1.2.3 Harmonic and general forcing excitation

Similarly to the SDOF (Sec. 1.1.2), here we focusing on the harmonic excitation with excitation load vector

$$\mathbf{f}(t) = \mathbf{F}\,\mathrm{e}^{\mathrm{i}\,\omega\,t}, \tag{1.163}$$

where $\mathbf{F}$ is the load amplitude vector, which is in general complex valued.

Modal damping

Different damping models have been discussed for the free response (Sec. 1.2.2); initially, we will study the forced response using the modal damping approach.

Using the *modal* or *principal* coordinates (see Eqs. (1.128) and (1.130)),

$$\mathbf{x}(t) = \mathbf{\Phi}\,\mathbf{q}(t), \tag{1.164}$$

and the modal damping (1.160) the equation of motion (1.101) changes similarly to Eq. (1.131); however, on the right-hand side the rth *modal force*

$$F_r = \mathbf{\Phi}_r^{\mathrm{T}}\,\mathbf{F} \tag{1.165}$$

is required:

$$\ddot{q}_r(t) + 2\,\zeta_r\,\omega_r\,\dot{q}_r(t) + \omega_r^2\,q_r(t) = F_r\,\mathrm{e}^{\mathrm{i}\,\omega\,t}, \tag{1.166}$$

where ω_r and ζ_r are the natural frequency and the modal damping for the rth mode ($r = 1, 2, \ldots, N$, where N is the number of degrees of freedom).

Similarly, to the SDOF case (1.34), the solution for the principal coordinates is defined as

$$q_r(t) = \alpha_r(\omega)\,F_r\,\mathrm{e}^{\mathrm{i}\,\omega\,t}, \tag{1.167}$$

where α_r is the receptance of mode r.

Taking into account the mass normalized modal matrix $\mathbf{\Phi}$, the transition from principal to physical coordinates is (1.129)

$$\begin{aligned}\mathbf{x}(t) &= \mathbf{\Phi}\,\mathbf{q}(t)\\ &= \sum_{r=1}^{N} \mathbf{\Phi}_r\, q_r(t)\\ &= \sum_{r=1}^{N} \mathbf{\Phi}_r\, \alpha_r(\omega)\, F_r\, \mathrm{e}^{\mathrm{i}\,\omega t}\\ &= \sum_{r=1}^{N} \mathbf{\Phi}_r\, \alpha_r(\omega)\, \mathbf{\Phi}_r^{\mathrm{T}}\, \mathbf{F}\, \mathrm{e}^{\mathrm{i}\,\omega t}\\ &= \sum_{r=1}^{N} \mathbf{\Phi}_r\, \mathbf{\Phi}_r^{\mathrm{T}}\, \mathbf{F}\, \alpha_r(\omega)\, \mathrm{e}^{\mathrm{i}\,\omega t}.\end{aligned} \tag{1.168}$$

Here it is reasonable to introduce the *frequency transfer function*[11], which relates the *input* or *excitation* denoted with index k and the *output* or *response* denoted with index j, $X_j = \alpha_{jk}\, F_k$:

$$\alpha_{jk}(\omega) = \sum_{r=1}^{N} \underbrace{\Phi_{jr}\, \Phi_{kr}}_{{}_rR_{jk}}\, \alpha_r(\omega), \tag{1.169}$$

where Φ_{jr} is the jth element of the rth mode $\mathbf{\Phi}_r$ (1.128), and ${}_rR_{jk}$ is the *modal constant*,

$${}_rR_{jk} = \Phi_{jr}\, \Phi_{kr}. \tag{1.170}$$

The FRF in physical coordinates $\alpha_{jk}(\omega)$ (1.169) is related to the FRF in principal/modal coordinates $\alpha_r(\omega)$ (1.35):

$$\alpha_{jk}(\omega) = \sum_{r}^{N} {}_rR_{jk}\, \alpha_r(\omega). \tag{1.171}$$

Fig. 1.14 shows the composition of a receptance, where different curves represent contribution of different mode shapes ${}_r\alpha_{jk}(\omega) = {}_rR_{jk}\, \alpha_r(\omega)$. The plot is in the logarithmic scale, so the absolute amplitudes are shown. The figure shows a 180° (or π) phase transition in the vicinity of the resonant frequency, which means that the receptance changes sign when passing the resonant frequency. As the FRF is composed of the sum of individual mode shapes and these have various signs depending on the modal constant and the phase varying with the frequency, they add or subtract from one another. At certain locations the modes

[11] Also called *FRF in physical coordinates.*

subtract to a minimal value, which is the point of an antiresonance. The antiresonance results in a 180° phase shift in the spectrum, because it indicates the border point where a different mode shape with an opposite sign is prevailing.

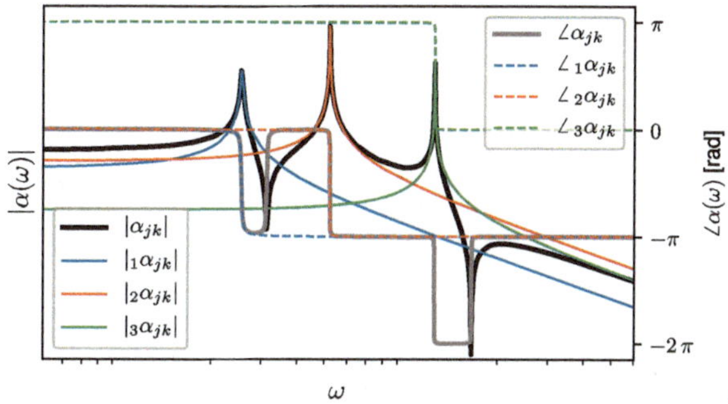

FIGURE 1.14 The composition of the frequency response function of an MDOF system.

Due to the orthogonality of the mode shapes $\boldsymbol{\Phi}$, the receptance matrix can also be written as

$$\boldsymbol{\alpha}(\omega) = \boldsymbol{\Phi} \begin{bmatrix} \ddots & & \\ & \alpha_r(\omega) & \\ & & \ddots \end{bmatrix} \boldsymbol{\Phi}^{\mathrm{T}}. \tag{1.172}$$

In case of viscous damping the modal receptance $\alpha_r(\omega)$ is defined by Eq. (1.126), and in case of hysteretic damping by Eq. (1.127).

The response in physical coordinates $\mathbf{x}(t)$ (1.168) can now be simplified to

$$\mathbf{x}(t) = \mathbf{X}\,\mathrm{e}^{\mathrm{i}\,\omega t} = \boldsymbol{\alpha}(\omega)\,\mathbf{F}\,\mathrm{e}^{\mathrm{i}\,\omega t} \tag{1.173}$$

in time-domain, which becomes in frequency-domain:

$$\mathbf{X}(\omega) = \boldsymbol{\alpha}(\omega)\,\mathbf{F}(\omega), \tag{1.174}$$

where ω is the angular frequency of the harmonic excitation $\mathbf{F}$ (in general, a complex quantity).

The *receptance transfer matrix* $\boldsymbol{\alpha}(\omega)$ is defined by elements α_{jk}, which relate the jth response to the kth excitation:

$$\boldsymbol{\alpha}(\omega) = \begin{bmatrix} \alpha_{11}(\omega) & \alpha_{12}(\omega) & \dots & \alpha_{1k}(\omega) & \dots & \alpha_{1N}(\omega) \\ \alpha_{21}(\omega) & \alpha_{22}(\omega) & \dots & \alpha_{2k}(\omega) & \dots & \alpha_{2N}(\omega) \\ \vdots & \vdots & \ddots & & & \vdots \\ \alpha_{j1}(\omega) & \alpha_{j2}(\omega) & \dots & \alpha_{jk}(\omega) & \dots & \alpha_{jN}(\omega) \\ \vdots & & & \vdots & \ddots & \vdots \\ \alpha_{N1}(\omega) & \alpha_{N2}(\omega) & \dots & \alpha_{Nk}(\omega) & \dots & \alpha_{NN}(\omega) \end{bmatrix}. \tag{1.175}$$

The receptance $\alpha_{jk}(\omega)$ is in general called the *point receptance*; if $j \neq k$, then it is called the *transfer receptance*; and if $j = k$, then it is called the *direct receptance*.

The excitation F_k at location k results in response at location j; similarly, the excitation F_j at location j results in response at location k. Due to reciprocity, we have

$$\alpha_{jk} = \frac{X_j}{F_k} = \alpha_{kj} = \frac{X_k}{F_j}. \tag{1.176}$$

Further, the modal constants are consistent with regard to the equations

$$\begin{aligned} {}_rR_{jk} &= \Phi_{jr}\,\Phi_{kr}, \\ {}_rR_{jj} &= \Phi_{jr}^2. \end{aligned} \tag{1.177}$$

Due to the reciprocity (1.176) and the modal constants consistency (1.177), the receptance matrix $\boldsymbol{\alpha}(\omega)$ is fully defined if one column or one row is measured.

If the modal constants matrix for the rth mode

$${}_r\mathbf{R} = \begin{bmatrix} \phi_{1r}\phi_{1r} & \phi_{1r}\phi_{2r} & \cdots & \phi_{1r}\phi_{kr} & \cdots & \phi_{1r}\phi_{Nr} \\ \phi_{2r}\phi_{1r} & \phi_{2r}\phi_{2r} & \cdots & \phi_{2r}\phi_{kr} & \cdots & \phi_{2r}\phi_{Nr} \\ \vdots & & \ddots & \vdots & & \vdots \\ \phi_{jr}\phi_{1r} & \phi_{jr}\phi_{2r} & \cdots & \phi_{jr}\phi_{kr} & \cdots & \phi_{jr}\phi_{Nr} \\ \vdots & & & \vdots & \ddots & \vdots \\ \phi_{Nr}\phi_{1r} & \phi_{Nr}\phi_{2r} & \cdots & \phi_{Nr}\phi_{kr} & \cdots & \phi_{Nr}\phi_{Nr} \end{bmatrix} \tag{1.178}$$

was experimentally identified (see Chapter 3), then the corresponding mass normalized mode shape is identified as

$$\boldsymbol{\Phi}_r = \frac{{}_r\mathbf{R}_j}{\phi_{jr}} = \frac{{}_r\mathbf{R}_k}{\phi_{kr}}, \tag{1.179}$$

where ${}_r\mathbf{R}_k$ is the kth column of the matrix ${}_r\mathbf{R}$.

Nonproportional viscous damping

In case the viscous damping is nonproportional, the state-space approach results in $2N$ complex conjugate eigenvalue pairs λ_r and λ_r^* and $2N$ complex conjugate eigenvector pairs $\mathbf{X}$ and $\mathbf{X}^*$ (see Eqs. (1.150)–(1.152)). If the mass normalized modal vectors are used, then the *point receptance* is defined as

$$\alpha_{jk}(\omega) = \sum_{r=1}^{N} \left(\frac{\Phi_{jr}\,\Phi_{kr}}{\mathrm{i}\,\omega - \lambda_r} + \frac{\Phi_{jr}^*\,\Phi_{kr}^*}{\mathrm{i}\,\omega - \lambda_r^*} \right), \tag{1.180}$$

and using the modal constant ${}_rR_{jk}$ (1.170) as

$$\alpha_{jk}(\omega) = \sum_{r=1}^{N} \left(\frac{{}_rR_{jk}}{\mathrm{i}\,\omega - \lambda_r} + \frac{{}_rR^*_{jk}}{\mathrm{i}\,\omega - \lambda^*_r} \right), \tag{1.181}$$

where λ_r is defined by Eq. (1.150).

The receptance of undamped structures is a real-valued function with positive and negative signs. When damping is considered, the receptance becomes complex. In the case where the system is not proportionally damped, the modal constants become complex as well (see Sec. 1.2.2). Fig. 1.15 shows a nonproportionally damped system. The complex nature of modal constants means that the antiresonances are less prominent, because the phase difference decreases the possibility for a complete subtraction; also, the mode shape is complex as well and no longer necessarily in-phase. When the mode shape is complex, different points on the structure are oscillating with different phases. As the points on the structure oscillate at different phases, it is common that all points are oscillating, resulting in a mode shape that does not have any nodes.

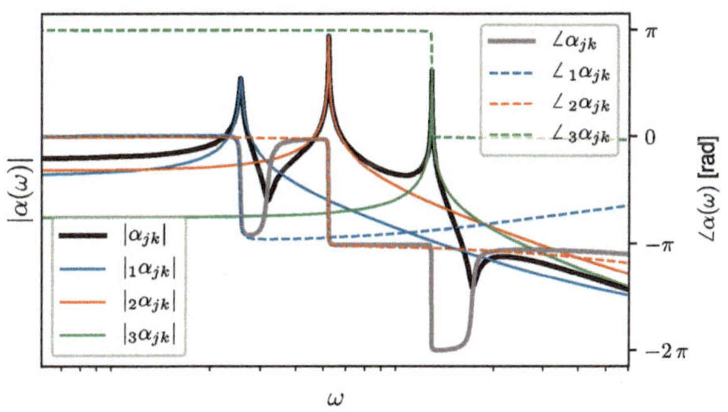

FIGURE 1.15 The composition of the frequency response function of a system with complex modal constants.

Relations between the spatial, modal, and response models

In structural dynamics, it is common to use the spatial, modal, or response model. All these models, together with other details (e.g., damping), have been previously discussed; for clarity, here we will briefly discuss and relate these models.

Spatial model

The multi-degree-of-freedom (MDOF) system with N degrees of freedom is described as a system of differential equations in matrix form (1.101):

$$\mathbf{M}\,\ddot{\mathbf{x}}(t) + \mathbf{C}\,\dot{\mathbf{x}}(t) + \mathbf{K}\,\mathbf{x}(t) = \mathbf{f}(t), \tag{1.182}$$

where the mass $\mathbf{M}$, damping $\mathbf{C}$ (or $\mathbf{D}$ for hysteretic damping), and stiffness $\mathbf{K}$ matrices of dimension $N \times N$ are bound by the (spatial) geometry and the material properties of the structure. If the dynamical system is defined by (1.182), then it is called the *spatial model*.

Modal model

In case of the generalized viscous damping model, Eq. (1.182) results in a set of $2N$ complex eigenvalues λ_r and $2N$ complex eigenvectors $\mathbf{\Psi}_r$ (1.148).The eigenvalues and eigenvectors appear in conjugate pairs,

$$\lambda_r = \lambda_{r+N}^*, \qquad \mathbf{\Psi}_r = \mathbf{\Psi}_{r+N}^*, \tag{1.183}$$

and form the *modal model*. The *eigenparameters* (1.183) are related to mode shapes, natural frequencies, and damping, and therefore the latter parameters are called *modal parameters*.

The eigenvectors are also known as *mode shapes* of the analyzed dynamical system. The eigenvectors/mode shapes are orthogonal (1.117) and therefore independent. In the spatial model the degrees-of-freedom are related to the spatial geometry of system (1.102), whereas in the modal model, they are related to the modal degrees of freedom (index r in Eq. (1.166)).

The modal model can be related to the spatial model using mass normalized mode shapes $\mathbf{\Phi}$; from the denominator in Eq. (1.159) we have

$$\mathbf{M} = \mathbf{\Phi}^{*-\mathbf{T}}\, \mathbf{\Phi}^{-1}, \tag{1.184}$$

where $\mathbf{\Phi}_r = \gamma_r\, \mathbf{\Psi}_r$ (1.121), and the scaling factor γ_r is defined by (1.123). The matrix $\mathbf{\Phi}$ has is of dimension $N \times N$ (not including the conjugate pairs), where the first matrix dimension spans the degrees-of-freedom, and the second one the eigenvalues.

Based on Eqs. (1.151)–(1.152) for the general *viscous* damped system, the stiffness $\mathbf{K}$ and damping $\mathbf{C}$ matrices are

$$\mathbf{K} = \mathbf{\Phi}^{*-\mathbf{T}} \begin{bmatrix} \ddots & & \\ & \omega_r^2 & \\ & & \ddots \end{bmatrix} \mathbf{\Phi}^{-1}, \tag{1.185}$$

$$\mathbf{C} = \mathbf{\Phi}^{*-\mathbf{T}} \begin{bmatrix} \ddots & & \\ & 2\,\omega_r\,\zeta_r & \\ & & \ddots \end{bmatrix} \mathbf{\Phi}^{-1}. \tag{1.186}$$

Based on Eq. (1.159), for the general *hysteretic* damped system, the stiffness $\mathbf{K}$ and damping $\mathbf{D}$ matrices are

$$\mathbf{K} + \mathrm{i}\,\mathbf{D} = \mathbf{\Phi}^{*-\mathrm{T}} \begin{bmatrix} \ddots & & \\ & \omega_r^2\,(1 + \mathrm{i}\,\eta_r) & \\ & & \ddots \end{bmatrix} \mathbf{\Phi}^{-1}. \tag{1.187}$$

The ratios of the eigenvector for a selected eigenvalue produce the dimensionless mode shape (the dimension is provided by the modal participation Q_r and provides a relation to excitation (1.162)).

Response model

The response in physical coordinates is related to the excitation force by Eq. (1.174):

$$\mathbf{X}(\omega) = \boldsymbol{\alpha}(\omega)\,\mathbf{F}(\omega). \tag{1.188}$$

The receptance matrix $\boldsymbol{\alpha}(\omega)$ (1.175) relates the jth response to the kth excitation (1.181):

$$\alpha_{jk}(\omega) = \sum_{r=1}^{N} \left(\frac{{}_rR_{jk}}{\mathrm{i}\,\omega - \lambda_r} + \frac{{}_rR_{jk}^*}{\mathrm{i}\,\omega - \lambda_r^*} \right), \tag{1.189}$$

where λ_r is defined by Eq. (1.150) in case of *viscous* and by Eq. (1.156) in case of *hysteretic* damping. The modal constant ${}_rR_{jk}$ is defined by mass normalized mode shapes (1.170):

$${}_rR_{jk} = \Phi_{jr}\,\Phi_{kr} \tag{1.190}$$

or by the dimensionless mode shapes $\mathbf{\Psi}_r$:

$${}_rR_{jk} = Q_r\,{}_r\Psi_j\,{}_r\Psi_k, \tag{1.191}$$

where Q_r is the scaling factor for the individual eigenvalue.[12]

The impulse response, as for the SDOF model, equals the inverse Fourier transform of the FRF and resembles the free-response (1.91):

$$h_{jk}(t) = \sum_{r=1}^{N} {}_rR_{jk}\,\mathrm{e}^{\lambda_r t} + {}_rR_{jk}^*\,\mathrm{e}^{\lambda_r^* t}. \tag{1.192}$$

The individual receptance $\alpha_{jk}(\omega)$ can be experimentally measured enabling the identification of modal parameters as shown with the previous equations. A

[12] The modal constant can be zero; if so, then the location it indicates is a node.

dynamical system of degree N is defined with a receptance matrix $\boldsymbol{\alpha}$ of dimension $N \times N$, which requires at least N measurements of receptance (1.177).

The receptance matrix $\boldsymbol{\alpha}$ defines the *response model* and relates the response to the modal model. Identification of modal parameters from the response model is called experimental modal analysis and is discussed in Chapter 3.

An overview of the relations between the spatial, modal, and response models is given in Table 1.3.

TABLE 1.3 Relations between different models.

Spatial model		Modal model		Response model
$\mathbf{M}, \mathbf{C}, \mathbf{D}, \mathbf{K}$	$\rightarrow$	Eigensystem solution		
viscous: (1.184)–(1.186) hysteretic: (1.184) and (1.187)	$\leftarrow$	$\boldsymbol{\Phi}_r, \lambda_r$	$\rightarrow$	(1.189)
		Modal identification	$\leftarrow$	$\boldsymbol{\alpha}(\omega)$

1.2.4 Base excitation

In Sec. 1.1.5 the two mass system presented in Fig. 1.9 was kinematically excited, which resulted in an SDOF system and the natural frequency of the system at $\omega_0 = \sqrt{k/m}$. In laboratory the kinematic excitation is possible with an electrodynamic shaker using a controller with a feedback loop. However, if the shaker is used without a controller (i.e., in an open loop, voltage driven), then the electromagnetic force inside the shaker is roughly proportional to the excitation signal voltage,[13] and the system is said to be *force* or *dynamically excited*. In the case of dynamic excitation the system has two degrees of freedom, and the equations of motions are obtained from the free-body diagram in Fig. 1.16. Neglecting the damping c, the equations of motion are

$$\begin{bmatrix} m & 0 \\ 0 & m_b \end{bmatrix} \begin{pmatrix} \ddot{x} \\ \ddot{y} \end{pmatrix} + \begin{bmatrix} k & -k \\ -k & k \end{bmatrix} \begin{pmatrix} x \\ y \end{pmatrix} = \begin{pmatrix} 0 \\ F(t) \end{pmatrix}. \tag{1.193}$$

The eigenvalues of the homogeneous part of Eq. (1.193) define the natural frequencies (1.110)

$$\omega_{0,1} = 0, \qquad \omega_{0,2} = \sqrt{\frac{k\,(m + m_b)}{m\, m_b}}. \tag{1.194}$$

[13] The force in an electromagnetic shaker is roughly proportional to the magnetic flux, which is roughly proportional to the drive current coming from the amplifier due to the excitation signal voltage.

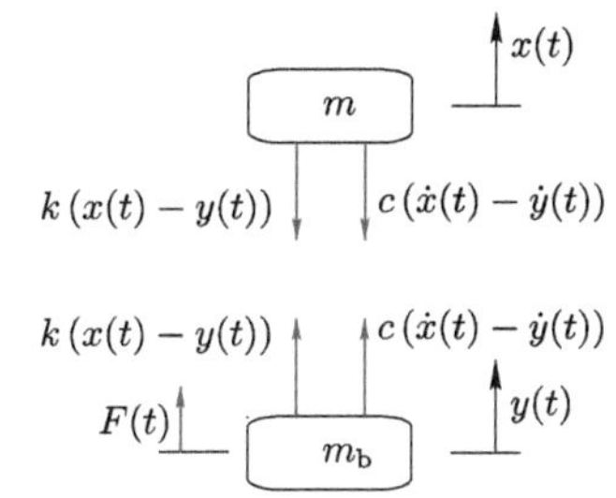

FIGURE 1.16 A two-degree-of-freedom model: free-body diagram.

If the natural frequency $\omega_{0,2}$ of the dynamically excited two-mass[14] system is compared to the natural frequency $\omega_0 = \sqrt{k/m}$ of the kinematically excited system, then it is clear that the difference is

$$\frac{\omega_{0,2}}{\omega_0} = \sqrt{\frac{m + m_b}{m_b}}. \tag{1.195}$$

The difference in dynamic vs. kinematic excitation was shown here on a two-mass system only; it is clear that the type of excitation significantly effects the natural dynamics of the researched structure. Not being aware of effect of the kinematic vs. dynamic excitation in numeric or experimental research can lead to wrong conclusions. A general MDOF base excited system is studied in detail by Česnik et al. [6].

1.2.5 Strain and stress response

In vibration fatigue, the damage is related to the stress or strain loads of the studied structure. The time-domain response $\mathbf{x}(t)$ (1.173) or the frequency-domain response $X(\omega)$ (1.174) needs to be related to the strain $\varepsilon(t)$ and stress $\sigma(t)$ response in time- and frequency-domains.

The strain tensor $\overline{\boldsymbol{\varepsilon}}$ in the time domain is defined as [7]

$$\overline{\boldsymbol{\varepsilon}} = \begin{bmatrix} \varepsilon_{xx} & \varepsilon_{xy} & \varepsilon_{xz} \\ \varepsilon_{yx} & \varepsilon_{yy} & \varepsilon_{yz} \\ \varepsilon_{yz} & \varepsilon_{zy} & \varepsilon_{zz} \end{bmatrix}. \tag{1.196}$$

The second-order symmetric strain tensor contains six independent components inside the body and reduces to three on the surface. For a linear elastic and isotropic material, the strain and stress tensors are symmetric and are typically

14 The natural frequency $\omega_{0,1}$ corresponds to the rigid-body motion when both $x(t)$ and $y(t)$ move in the same direction and $x(t) - y(t) = 0$.

defined in the Voigt vector notation[15]:

$$\boldsymbol{\varepsilon} = [\varepsilon_{xx}, \varepsilon_{yy}, \varepsilon_{zz}, 2\,\varepsilon_{xy}, 2\,\varepsilon_{xz}, 2\,\varepsilon_{yz}]^{\mathrm{T}}. \tag{1.197}$$

An alternative nomenclature for the strain tensor is

$$\mathbf{e} = [e_1, e_2, e_3, e_4, e_5, e_6]^{\mathrm{T}}. \tag{1.198}$$

The infinitesimal strain tensor $\overline{\boldsymbol{\varepsilon}}$ is defined from *small* displacement field **u** [8]:

$$\overline{\boldsymbol{\varepsilon}} = \mathrm{D}\{\mathbf{u}\}, \tag{1.199}$$

where $\mathrm{D}\{\cdot\}$ is the linear differential operator defined as

$$\mathrm{D}\{\cdot\} = \frac{1}{2}\left(\nabla + \nabla^{\mathrm{T}}\right) \tag{1.200}$$

and ∇ is the nabla-Hamiltonian (gradient) operator.

The linear spatial differential operator in the Voight notation is

$$\mathrm{D}\{\cdot\} = \begin{bmatrix} \partial/\partial x & 0 & 0 \\ 0 & \partial/\partial y & 0 \\ 0 & 0 & \partial/\partial z \\ \partial/\partial y & \partial/\partial x & 0 \\ \partial/\partial z & 0 & \partial/\partial x \\ 0 & \partial/\partial z & \partial/\partial y \end{bmatrix}. \tag{1.201}$$

The stress is defined as

$$\overline{\boldsymbol{\sigma}} = \begin{bmatrix} \sigma_{xx} & \sigma_{xy} & \sigma_{xz} \\ \sigma_{yx} & \sigma_{yy} & \sigma_{yz} \\ \sigma_{yz} & \sigma_{zy} & \sigma_{zz} \end{bmatrix} \tag{1.202}$$

or, in the Voigt vector notation,

$$\boldsymbol{\sigma} = [\sigma_{xx}, \sigma_{yy}, \sigma_{zz}, \sigma_{xy}, \sigma_{xz}, \sigma_{yz}]^{\mathrm{T}}. \tag{1.203}$$

An alternative nomenclature for the stress tensor, which will later be used for clarity, is

$$\mathbf{s} = [s_1, s_2, s_3, s_4, s_5, s_6]^{\mathrm{T}}. \tag{1.204}$$

15 Other books might have a different order of components; in this book, we always use the order used here.

The stress can be related to strain [9]:

$$\begin{bmatrix}\sigma_{xx}\\ \sigma_{yy}\\ \sigma_{zz}\\ \sigma_{xy}\\ \sigma_{xz}\\ \sigma_{yz}\end{bmatrix}=\begin{bmatrix}C_{11} & C_{12} & C_{12} & 0 & 0 & 0\\ C_{12} & C_{11} & C_{12} & 0 & 0 & 0\\ C_{12} & C_{12} & C_{11} & 0 & 0 & 0\\ 0 & 0 & 0 & \frac{C_{11}-C_{12}}{2} & 0 & 0\\ 0 & 0 & 0 & 0 & \frac{C_{11}-C_{12}}{2} & 0\\ 0 & 0 & 0 & 0 & 0 & \frac{C_{11}-C_{12}}{2}\end{bmatrix}\begin{bmatrix}\varepsilon_{xx}\\ \varepsilon_{yy}\\ \varepsilon_{zz}\\ 2\,\varepsilon_{xy}\\ 2\,\varepsilon_{xz}\\ 2\,\varepsilon_{yz}\end{bmatrix}, \tag{1.205}$$

where

$$C_{11}=\frac{E\,(1-\nu)}{(1+\nu)\,(1-2\,\nu)},\qquad C_{12}=\frac{E\,\nu}{(1+\nu)\,(1-2\,\nu)},$$
$$\frac{C_{11}-C_{12}}{2}=\frac{E}{2\,(1+\nu)}; \tag{1.206}$$

E is the *Young modulus of elasticity* (ratio of the axial stress to the axial strain), G is the *shear modulus of elasticity*, and ν is the *Poisson ratio* (ratio of the transverse contraction to the axial elongation in simple tension load). Shear and Young's moduli of elasticity are related by

$$G=\frac{E}{2\,(1+\nu)}. \tag{1.207}$$

Eq. (1.205) can be shortly written as

$$\boldsymbol{\sigma}=\mathbf{C}\,\boldsymbol{\varepsilon}, \tag{1.208}$$

where **C** is the elasticity matrix.

The relation from stress to strain is defined as

$$\begin{bmatrix}\varepsilon_{xx}\\ \varepsilon_{yy}\\ \varepsilon_{zz}\\ 2\,\varepsilon_{xy}\\ 2\,\varepsilon_{xz}\\ 2\,\varepsilon_{yz}\end{bmatrix}=\begin{bmatrix}V_{11} & V_{12} & V_{12} & 0 & 0 & 0\\ V_{12} & V_{11} & V_{12} & 0 & 0 & 0\\ V_{12} & V_{12} & V_{11} & 0 & 0 & 0\\ 0 & 0 & 0 & 2\,(V_{11}-V_{12}) & 0 & 0\\ 0 & 0 & 0 & 0 & 2\,(V_{11}-V_{12}) & 0\\ 0 & 0 & 0 & 0 & 0 & 2\,(V_{11}-V_{12})\end{bmatrix}\begin{bmatrix}\sigma_{xx}\\ \sigma_{yy}\\ \sigma_{zz}\\ \sigma_{xy}\\ \sigma_{xz}\\ \sigma_{yz}\end{bmatrix}, \tag{1.209}$$

where

$$V_{11}=\frac{1}{E},\qquad V_{12}=\frac{-\nu}{E},\qquad 2\,(V_{11}-V_{12})=\frac{2\,(1+\nu)}{E}. \tag{1.210}$$

Eq. (1.209) can be shortly written as

$$\boldsymbol{\varepsilon} = \mathbf{V}\,\boldsymbol{\sigma}, \tag{1.211}$$

where $\mathbf{V}$ is the compliance matrix; it can be shown that $\mathbf{V} = \mathbf{C}^{-1}$.

For the stress tensor $\boldsymbol{\sigma}$ given in an orthogonal coordinate system, the *principal stresses* can be found via coordinate transformations. The principal stresses σ_1, σ_2, and σ_3 are acting on orthogonal planes that are free of shear stresses. For a given stress tensor $\boldsymbol{\sigma}$, the principal stresses are defined by finding the roots of the cubic equation [7,10]

$$\sigma^3 - \sigma^2\, I_1 + \sigma\, I_2 - I_3 = 0, \tag{1.212}$$

where

$$\begin{aligned} I_1 &= \sigma_{xx} + \sigma_{yy} + \sigma_{zz} \\ &= \sigma_1 + \sigma_2 + \sigma_3, \\ I_2 &= \sigma_{xx}\,\sigma_{yy} + \sigma_{yy}\,\sigma_{zz} + \sigma_{xx}\,\sigma_{zz} - \sigma_{xy}^2 - \sigma_{yz}^2 - \sigma_{xz}^2 \\ &= (\sigma_1\,\sigma_2 + \sigma_2\,\sigma_3 + \sigma_3\,\sigma_1)\,, \\ I_3 &= \sigma_{xx}\,\sigma_{yy}\,\sigma_{zz} + 2\,\sigma_{xy}\,\sigma_{yz}\,\sigma_{zx} - \sigma_{xy}^2\,\sigma_{zz} - \sigma_{yz}^2\,\sigma_{xx} - \sigma_{xz}^2\,\sigma_{yy} \\ &= \sigma_1\,\sigma_2\,\sigma_3, \end{aligned} \tag{1.213}$$

where the stress invariants are defined in the general xyz coordinate system ($\sigma_{xx}, \sigma_{xy}, \ldots$) and the principal coordinates (σ_1, σ_2, σ_3).

The principal stresses can also be defined by finding the eigenvalues of $\overline{\boldsymbol{\sigma}}$ (1.202), where the eigenvectors define the principal stress directions.

Strain modal response

Up to now, the displacement response and the *displacement mode shapes* (DMS) have been discussed. In vibration fatigue the damage is related to the stress or strain loads of the researched structure, and therefore the *strain response* and the *strain mode shapes* (SMS) will be discussed next.

To obtain the rth mass-normalized SMS $\boldsymbol{\Phi}_r^{e_i}$ for the selected component i (1.198), the linear differential operator $\mathrm{D}\{\cdot\}$ (1.201) is applied to the mass-normalized DMS $\boldsymbol{\Phi}_r$ for this component only (1.123) [8,11–13]:

$$\boldsymbol{\Phi}_r^{e_i} = \mathrm{D}_i\{\boldsymbol{\Phi}_r\}. \tag{1.214}$$

If the operator $\mathrm{D}\{\cdot\}$ is applied to the time-domain response $\mathbf{x}(t)$ (1.168), only the mass normalized modal matrix $\boldsymbol{\Phi}$ needs to be differentiated. It follows that the amplitude of the ith component of the strain steady-state response $\mathbf{X}_{e_i}(\omega)$

is (1.172), (1.174) [8]

$$\begin{aligned}\mathbf{X}_{e_i}(\omega) &= \mathrm{D}_i\{\boldsymbol{\Phi}\}\begin{bmatrix}\ddots & & \\ & \alpha_r(\omega) & \\ & & \ddots\end{bmatrix}\boldsymbol{\Phi}^{\mathrm{T}}\,\mathbf{F}(\omega)\\ &= \boldsymbol{\Phi}^{e_i}\begin{bmatrix}\ddots & & \\ & \alpha_r(\omega) & \\ & & \ddots\end{bmatrix}\boldsymbol{\Phi}^{\mathrm{T}}\,\mathbf{F}(\omega)\\ &= \mathbf{H}_{e_i f}(\omega)\,\mathbf{F}(\omega),\end{aligned} \tag{1.215}$$

where hysteretic damping was assumed, and $\mathbf{H}_{e_i f}(\omega)$ is the strain frequency response function matrix from the excitation force to the response strain component e_i [13]:

$$\mathbf{H}_{e_i f}(\omega) = \sum_{r=1}^{N} \frac{{}_r\mathbf{R}^{e_i}}{\omega_r^2 - \omega^2 + \mathrm{i}\,\eta_r\,\omega_r^2}; \tag{1.216}$$

${}_r\mathbf{R}^{e_i}$ is the e_i *strain modal constants matrix component* of the rth mode defined as

$${}_r\mathbf{R}^{e_i} = \begin{bmatrix} \phi_{1r}^{e_i}\,\phi_{1r} & \phi_{1r}^{e_i}\,\phi_{2r} & \cdots & \phi_{1r}^{e_i}\,\phi_{kr} & \cdots & \phi_{1r}^{e_i}\,\phi_{N^{\mathrm{d}}r} \\ \phi_{2r}^{e_i}\,\phi_{1r} & \phi_{2r}^{e_i}\,\phi_{2r} & \cdots & \phi_{2r}^{e_i}\,\phi_{kr} & \cdots & \phi_{2r}^{e_i}\,\phi_{N^{\mathrm{d}}r} \\ \vdots & & \ddots & \vdots & & \vdots \\ \phi_{jr}^{e_i}\,\phi_{1r} & \phi_{jr}^{e_i}\,\phi_{2r} & \cdots & \phi_{jr}^{e_i}\,\phi_{kr} & \cdots & \phi_{jr}^{e_i}\,\phi_{N^{\mathrm{d}}r} \\ \vdots & & & \vdots & \ddots & \vdots \\ \phi_{N^{\mathrm{s}}r}^{e_i}\,\phi_{1r} & \phi_{N^{\mathrm{s}}r}^{e_i}\,\phi_{2r} & \cdots & \phi_{N^{\mathrm{s}}r}^{e_i}\,\phi_{kr} & \cdots & \phi_{N^{\mathrm{s}}r}^{e_i}\,\phi_{N^{\mathrm{d}}r} \end{bmatrix}, \tag{1.217}$$

where $\phi_{jr}^{e_i}$ and ϕ_{kr} are the elements of strain mass normalized mode shape $\boldsymbol{\Phi}_r^{e_i}$ and the displacement mass normalized mode shape $\boldsymbol{\Phi}_r$, respectively, N^{s} and N^{d} are the sizes of $\boldsymbol{\Phi}_r^{e_i}$ and $\boldsymbol{\Phi}_r$, respectively. In general, $\mathbf{H}_{e_i f}$ is not a square matrix [13], and the strain modal constant matrix is not symmetric (${}_rR_{jk}^{e_i} \neq {}_rR_{kj}^{e_i}$); therefore $H_{e_i f,jk} \neq H_{e_i f,kj}$, and the mass normalization, which was in deflection mode shapes quite straightforward (1.179), requires the combination of the deflection and strain measurements.

From the jth row of the strain modal constants matrix $\left({}_r\mathbf{R}^{e_i}\right)_j^{\mathrm{T}}$ the mass normalized displacement mode shape $\boldsymbol{\Phi}_r$ is defined as

$$\boldsymbol{\Phi}_r = \frac{\left({}_r\mathbf{R}^{e_i}\right)_j^{\mathrm{T}}}{\phi_{jr}^{e_i}}. \tag{1.218}$$

Similarly, from the kth column of the strain modal constant $\left({}_r\mathbf{R}^{e_i}\right)_k$ the mass normalized strain mode shape component $\mathbf{\Phi}_r^{e_i}$ is defined as

$$\mathbf{\Phi}_r^{e_i} = \frac{{}_r\mathbf{R}_k^{e_i}}{\phi_{kr}}. \tag{1.219}$$

Identifying displacement and strain mass normalized mode shapes (1.218)–(1.219) requires the measurement of displacement and strain response; at least one row and at least one column of the strain modal constants matrix needs to be obtained; for details on mass normalization of strain mode shapes, we refer to [12,14].

It is important to note here that if we use the finite element method, then the vibration fatigue analysis usually starts with the numerically obtained DMS $\mathbf{\Phi}_r$ and SMS $\mathbf{\Phi}_r^{e_i}$ which define the FRF from force to strain (1.216).

Stress modal response

The infinitesimal stress tensor $\overline{\boldsymbol{\sigma}}$ and the infinitesimal displacement field $\mathbf{u}$ are related as

$$\overline{\boldsymbol{\sigma}} = \mathbf{C}\,\mathrm{D}\{\mathbf{u}\} = \mathrm{S}\{\mathbf{u}\}, \tag{1.220}$$

where D is the linear differential operator (1.200), and C is the elasticity matrix (1.208); $\mathrm{S}\{\cdot\}$ is the linear operator from the infinitesimal displacement field to infinitesimal stress.

To obtain the rth mass-normalized *stress mode shape* (SsMS) $\mathbf{\Phi}_r^{s_i}$ for the selected component i, the operator S_i is applied to the mass-normalized DMS $\mathbf{\Phi}_r$ (1.123) for this component only:

$$\mathbf{\Phi}_r^{s_i} = \mathrm{S}_i\{\mathbf{\Phi}_r\}. \tag{1.221}$$

The amplitude of the ith component of the stress steady-state response $\mathbf{X}_{s_i}(\omega)$ is defined similarly as for the strain (1.215):

$$\begin{aligned}
\mathbf{X}_{s_i}(\omega) &= \mathrm{S}_i\{\mathbf{\Phi}\}\begin{bmatrix}\ddots & & \\ & \alpha_r(\omega) & \\ & & \ddots\end{bmatrix}\mathbf{\Phi}^{\mathrm{T}}\,\mathbf{F}(\omega) \\
&= \mathbf{\Phi}^{s_i}\begin{bmatrix}\ddots & & \\ & \alpha_r(\omega) & \\ & & \ddots\end{bmatrix}\mathbf{\Phi}^{\mathrm{T}}\,\mathbf{F}(\omega) \\
&= \mathbf{H}_{s_i f}(\omega)\,\mathbf{F}(\omega),
\end{aligned} \tag{1.222}$$

where hysteretic damping is assumed, and $\mathbf{H}_{s_i f}(\omega)$ is the stress frequency response function matrix from the excitation force to the response stress component s_i:

$$\mathbf{H}_{s_i f}(\omega) = \sum_{r=1}^{N} \frac{{}_r\mathbf{R}^{s_i}}{\omega_r^2 - \omega^2 + \mathrm{i}\,\eta_r\,\omega_r^2}; \tag{1.223}$$

${}_r\mathbf{R}^{s_i}$ is the s_i component of the *stress modal constants matrix* of the rth mode defined similarly to (1.217), where

$${}_r\mathbf{R}_{jk}^{s_i} = \phi_{jr}^{s_i}\,\phi_{kr} \tag{1.224}$$

is the jk element of the rth mode.

It is again important to note that the SsMS $\mathbf{\Phi}_r^{s_i}$ are usually directly obtained from FEM models (with partial experimental validation).

1.2.6 Incomplete models

When measuring FRFs, only a limited frequency span is measured, the higher frequencies are limited with the sampling frequency and the dynamic response of the used sensors, whereas the lower frequencies are limited with the frequency resolution and the charge amplifiers in case of using piezoelectric accelerometers, the latter cutting the low-frequency content. Additional limitations are set by the excitation either because of the large motion necessary for the lower frequencies or because of the inability of higher-frequency excitation due to the large inertia and low rigidity of the excitation device.

The limited frequency range offers only an incomplete model of the true structural dynamics. For the case of the receptance,

$$\alpha(\omega) = \alpha_L(\omega) + \alpha_{in}(\omega) + \alpha_U(\omega), \tag{1.225}$$

where $\alpha_L(\omega)$ represents the residual of the modes below the lowest measured frequency $\omega \le \omega_{r_L}$,

$$\alpha_L(\omega) = \sum_{r=1}^{r_L} \left(\frac{{}_rR_{jk}}{\mathrm{i}\,\omega - \lambda_r} + \frac{{}_rR_{jk}^*}{\mathrm{i}\,\omega - \lambda_r^*} \right), \tag{1.226}$$

$\alpha_{in}(\omega)$ represents the dynamic response inside the measured frequency range $\omega_{r_L} < \omega \le \omega_{r_U}$,

$$\alpha_{in}(\omega) = \sum_{r=r_L+1}^{r_U} \left(\frac{{}_rR_{jk}}{\mathrm{i}\,\omega - \lambda_r} + \frac{{}_rR_{jk}^*}{\mathrm{i}\,\omega - \lambda_r^*} \right), \tag{1.227}$$

and $\alpha_U(\omega)$ represents the residual of the modes above the measured frequency range $\omega_{r_U} < \omega$,

$$\alpha_U(\omega) = \sum_{r=r_U+1}^{N} \left(\frac{{}_rR_{jk}}{\mathrm{i}\,\omega - \lambda_r} + \frac{{}_rR_{jk}^*}{\mathrm{i}\,\omega - \lambda_r^*} \right). \tag{1.228}$$

The limited frequency range can cause problems in the process of system identification, because the modes outside the measured frequency range influence the FRF of the measured range (Fig. 1.17). The black line in Fig. 1.17 indicates the measured FRF, which equals the true FRF, only that it is cut to a limited frequency range, assuming no aliasing and leakage problems. The true FRF differs from the FRF composed of only the modes inside the measured range α_{in}; this is because the residuals from the modes below $\alpha_L(\omega)$ and above $\alpha_U(\omega)$ the measured range influence the response elsewhere as well; see Fig. 1.17.

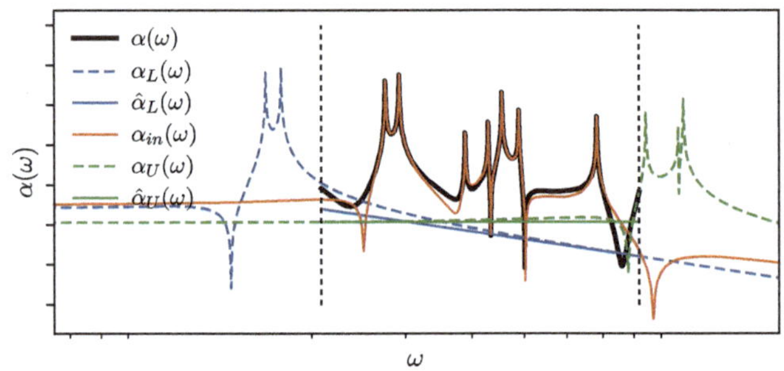

FIGURE 1.17 A limited measured frequency range and the influence of the outer modes. Vertical dotted lines denote ω_L and ω_U.

The residues above the resonant mode can be approximated by a mass characteristic [4]:

$$m\ddot{x} = f \rightarrow \alpha(\omega) = -\frac{1}{\omega^2 m}. \tag{1.229}$$

Due to this, the receptance in the frequency range (ω_L, ω_U) can be approximated as

$$\hat{\alpha}_L(\omega) \approx -\frac{R_L}{\omega^2}, \tag{1.230}$$

where R_L is the lower residual.

Similarly, the residues below the resonant mode can be approximated by a stiffness characteristic:

$$k\,x = f \rightarrow \alpha(\omega) = \frac{1}{k}. \tag{1.231}$$

Due to this, the receptance in the frequency range (ω_L, ω_U) can be approximated as

$$\hat{\alpha}_U(\omega) \approx R_U, \tag{1.232}$$

where R_U is the upper residual.

Finally, the incomplete model (1.225) can be approximated with

$$\alpha(\omega) \approx \sum_{r=r_L+1}^{r_U} \left(\frac{{}_rR_{jk}}{\mathrm{i}\,\omega - \lambda_r} + \frac{{}_rR_{jk}^*}{\mathrm{i}\,\omega - \lambda_r^*} \right) - \frac{R_L}{\omega^2} + R_U, \tag{1.233}$$

where the upper R_U and lower R_L residuals are identified with the modal identification method; see Sec. 3.3.4. The approximated receptances $\hat{\alpha}_L(\omega)$ and $\hat{\alpha}_U(\omega)$ are shown in Fig. 1.17.

Chapter 2

Signal processing

Signal processing deals with the analysis of processes that can in general be *deterministic* or *random*. A deterministic process is a process that can (with reasonable accuracy) be described with an explicit mathematical relation. A random process is a process than cannot be described with an explicit mathematical relation, but requires probability statements and statistical averages.

In this chapter, we give a general introduction required for the following chapters. Greater details can be found in books by Bendat and Piersol [15], Shin and Hammond [16], Newland [17], and Rice [18].

2.1 Deterministic process

Deterministic data from a deterministic process can be unambiguously deduced through mathematical relationships and conditions. Values within the sequence, such as a time history, can be successfully extrapolated or predicted, at least within the tolerable error margin. When considering, for instance, the measured data from a deterministic dynamic system, it is expected that, under equal conditions, the observed repeated system response will always be equal, as is, for example, for a mathematical pendulum released from the same height.

2.1.1 Fourier integral transform

The vibrations of linear structures are of harmonic nature and therefore typically analyzed in the frequency domain. The Fourier transform is commonly used to transform data to and from the frequency domain and was already defined by Eqs. (1.70), and (1.71):

$$X(\omega) = \int_{-\infty}^{\infty} x(t)\,\mathrm{e}^{-\mathrm{i}\,\omega\,t}\,\mathrm{d}t, \qquad x(t) = \frac{1}{2\pi}\int_{-\infty}^{\infty} X(\omega)\,\mathrm{e}^{\mathrm{i}\,\omega\,t}\,\mathrm{d}\omega, \tag{2.1}$$

where the Fourier integral transform is denoted with $\mathcal{F}\{\}$, and ω is the angular frequency. The inverse Fourier transform is denoted with $\mathcal{F}^{-1}\{\}$.

The numerical methods typically use the *frequency in seconds* (i.e., Hz); in this case the normalized, *unitary* definition is given by the Fourier pair (Eqs. (1.72)–(1.73)):

$$X(f) = \int_{-\infty}^{\infty} x(t)\,\mathrm{e}^{-\mathrm{i}\,2\pi\,f\,t}\,\mathrm{d}t, \qquad x(t) = \int_{-\infty}^{\infty} X(f)\,\mathrm{e}^{\mathrm{i}\,2\pi\,f\,t}\,\mathrm{d}f. \tag{2.2}$$

Vibration Fatigue by Spectral Methods. https://doi.org/10.1016/B978-0-12-822190-7.00009-8

2.1.2 Discrete Fourier transform

When signal is measured for finite amount of time T at discrete time instances $n\,\Delta t$, where Δt defines the sampling frequency $f_s = 1/\Delta t$, and $n = 0, 1, \ldots, N-1$ (see Fig. 2.1); N is the number of samples and is related to the sampling time $T = \Delta t\, N$.

At discrete-valued $x_n = x(n\,\Delta t)$ the integral (2.2) is discretized into the sum

$$X_k = \sum_{n=0}^{N-1} x_n\, \mathrm{e}^{\frac{-\mathrm{i}\,2\pi}{N}\, n\, k}, \tag{2.3}$$

where the spectral components are also discrete: $X_k = X(k\,\Delta f)$ and $k = 0, 1, \ldots, N-1$. The frequency resolution Δf is defined as

$$\Delta f = \frac{1}{T} = \frac{1}{\Delta t\, N}. \tag{2.4}$$

Eq. (2.3) defines the discrete Fourier transform (DFT), whereas the inverse discrete Fourier transform (IDFT) is defined as

$$x_n = \frac{1}{N} \sum_{k=0}^{N-1} X_k\, \mathrm{e}^{\frac{\mathrm{i}\,2\pi}{N}\, n\, k}. \tag{2.5}$$

The computing complexity of DFT is $O(n^2)$ and typically relies on the *fast Fourier transform* (FFT) algorithm, which has a significantly lower computing complexity $O(n\,\log(n))$ [19].

2.1.3 Energy spectra and Parseval's theorem

The energy in time domain is frequently related to the square of a known value, for example, displacement in potential energy, velocity in kinetic energy, voltage in the energy dissipation of a resistor. $x(t)$ is the value related to the instantaneous power as

$$\frac{1}{T} \int_0^T x(t)^2\, \mathrm{d}t. \tag{2.6}$$

It can be shown [16,17] that the energy is related to the Fourier transform of $x(t)$:

$$\int_{-\infty}^{+\infty} x(t)^2\, \mathrm{d}t = \int_{-\infty}^{+\infty} |X(f)|^2\, \mathrm{d}f = \frac{1}{2\pi} \int_{-\infty}^{+\infty} |X(\omega)|^2\, \mathrm{d}\omega, \tag{2.7}$$

where $X(\omega) = \mathcal{F}\{x(t)\}$. Eq. (2.7) is called the *Parseval theorem*. It is important as it relates the time- and frequency-domain energy.

2.1.4 Properties of the Fourier transform

In mathematical manipulations of the Fourier transform, we frequently relate to the properties of the Fourier transform:

$$\mathcal{F}\{a\,x(t)+b\,y(t)\}=a\,\mathcal{F}\{x(t)\}+b\,\mathcal{F}\{y(t)\} \quad \text{(linearity)}, \tag{2.8}$$

$$\mathcal{F}\{x(t\pm t_0)\}=X(\omega)\,\mathrm{e}^{\pm\mathrm{i}\,\omega\,t_0} \quad \text{(time shift)}, \tag{2.9}$$

$$\mathcal{F}^{-1}\{X(\mathrm{i}\,\omega\pm\omega_0)\}=x(t)\,\mathrm{e}^{\mp\mathrm{i}\,\omega_0\,t} \quad \text{(frequency shift)}, \tag{2.10}$$

$$\mathcal{F}\{x(-t)\}=X(-\omega) \quad \text{(time reversal)}, \tag{2.11}$$

$$\mathcal{F}\{x(a\,t)\}=\frac{1}{a}\,X\left(\frac{\omega}{a}\right) \quad \text{(time scaling)}, \tag{2.12}$$

$$\mathcal{F}\left\{\frac{\mathrm{d}^n}{\mathrm{d}t^n}\,x(t)\right\}=(\mathrm{i}\,\omega)^n\,X(\omega) \quad \text{(differentiation)}, \tag{2.13}$$

$$\mathcal{F}\left\{\int x(t)\,\mathrm{d}t\right\}=\frac{1}{\mathrm{i}\,\omega}\,X(\omega) \quad \text{(integration)}. \tag{2.14}$$

Real-valued function

A Fourier transform of a real valued function[1] $x(t)$ produces a Hermitian function $X(\omega)$. A Hermitian function is conjugate symmetric:

$$X(\omega)=X^*(-\omega). \tag{2.15}$$

$X(\omega)$ is frequently also called the *two-sided* Fourier transform, whereas in numerical code (in case of a real signal $x(t)$) the *one-sided* version is typically used:

$$\tilde{X}(\omega)=\begin{cases} 2\,X(\omega), & \omega>0,\\ X(\omega), & \omega=0,\\ 0, & \omega<0. \end{cases} \tag{2.16}$$

Convolution in the frequency

In the following sections, we present the cases of multiplying a signal $x(t)$ by another function in the time domain $w(t)$ with intent on observing the effects in the frequency domain and vice versa. The Fourier transform of the product

[1] In mechanical engineering, most of time-domain functions are real-valued.

of two functions in the time-domain equals the convolution in the frequency-domain:

$$\mathcal{F}\{x(t) \cdot w(t)\} = \mathcal{F}\{x(t)\} * \mathcal{F}\{w(t)\}, \tag{2.17}$$

where the convolution of two functions for a causal system is defined by Eq. (1.84), and for a noncausal system, by Eq. (1.86). The relation can also be reversed: the Fourier transform of the time-domain convolution equals the multiplication in the frequency domain:

$$\mathcal{F}\{x(t) * w(t)\} = \mathcal{F}\{x(t)\} \cdot \mathcal{F}\{w(t)\}. \tag{2.18}$$

2.1.5 Sampling, aliasing, and window functions

In this section, we discuss some practical aspects of measuring physical quantities; for details, see [20,16].

Zero-padding

In real measurements the sampling period T is finite, and consequently the frequency resolution is $1/T$ (see Sec. 2.1.2). Due to the limited sampling period T, the frequency resolution is also limited and often insufficient. The problem can be partially mitigated by adding zeros to the signal, thereby extending it and improving the frequency resolution. The concept is known as zero-padding. Note, however, that the improved resolution is not derived from the actual measurement and does not necessarily indicate the true state.

Aliasing

The discrete sampling of a signal $x(t)$ equals the product of this continuous signal and a sampling signal $s(t)$:

$$x_s(t) = x(t)\, s(t). \tag{2.19}$$

The sampling signal is comprised of a train of pulses indicating the sampling instances with the sampling period $\Delta t = 1/f_\mathrm{s}$, where f_s is the sampling frequency (Fig. 2.1). The result of the product is a discrete signal $x_s(t)$.

Assuming that the pulses are infinitely short (train of Dirac impulses), the sampling function $s(t)$ equals a function $S(\omega)$ in the frequency domain, composed of a sequence of Dirac functions as well, with these spaced apart by the sampling frequency f_s (ω_s); see Fig. 2.2B.

The discrete sampled signal in the frequency domain $X_s(\omega)$ equals the convolution of the signal with the sampling function in frequency-domain:

$$X_s(\omega) = X(\omega) * S(\omega). \tag{2.20}$$

As $S(\omega)$ is a train of Dirac impulses, convolution (2.20) produces repeated signal spectra at every multiple of the sampling frequency. The measured spectrum $X_s(\omega)$ has a limited range; see Fig. 2.2A. If the measured spectrum X_s

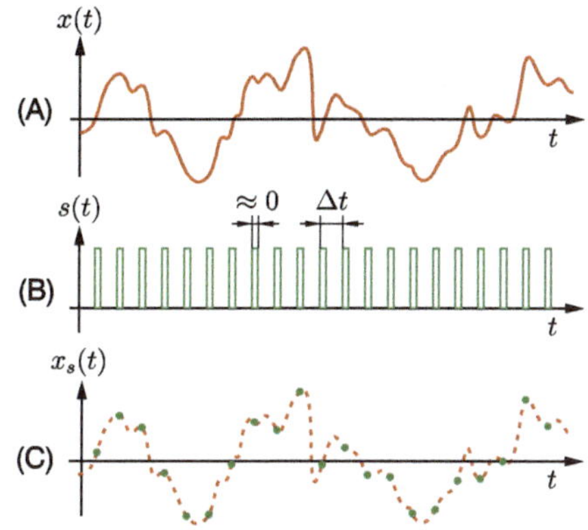

FIGURE 2.1 (A) The continuous signal $x(t)$, (B) the sampling function $s(t)$, and (C) the sampled discrete signal $x_S(t)$.

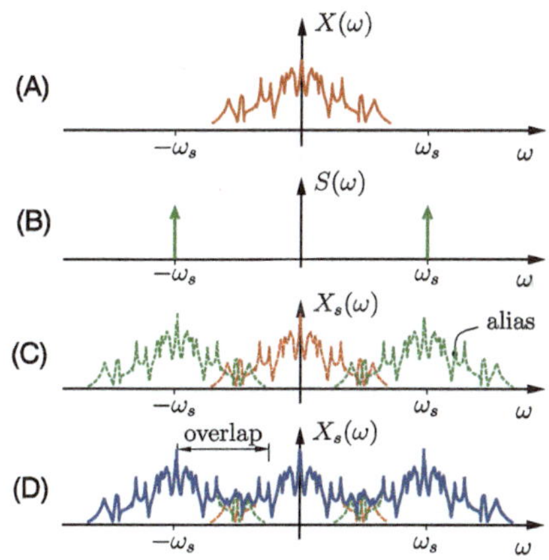

FIGURE 2.2 (A) The continuous signal $X(\omega)$, (B) the sampling function $S(\omega)$, (C) aliasing, and (D) the measured spectrum $X_S(\omega)$.

spans outside the measured range of frequencies, then it will be aliased as shown in Figs. 2.2C and D. Aliasing is seen in the time domain as overlapping harmonics with different frequencies, all passing the same sampled instances; see Fig. 2.3.

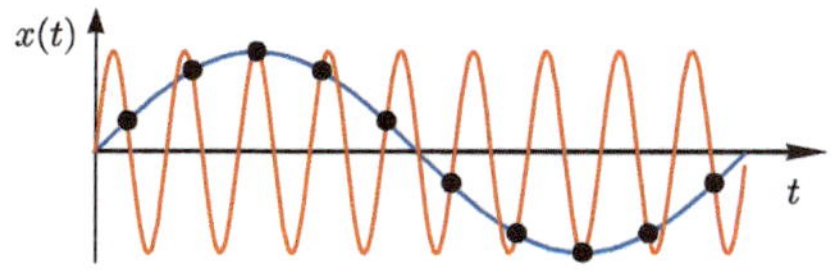

FIGURE 2.3 Aliasing in time domain.

To avoid aliasing, the sampling frequency should be at least twice the highest frequency in the measured signal f_{max} (the Nyquist–Shannon sampling theorem):

$$f_s > 2\, f_{\text{max}}. \tag{2.21}$$

The measured dynamic response may exceed the measured frequency range; to ensure no aliasing occurs, low-pass filters are used prior to the analog to digital conversion, or only a selected frequency range is excited when using a shaker to excite the structure.

Window functions

The sampling period is always finite, and therefore the discrete Fourier transform is limited to a finite interval. By measuring for a finite period, the measured signal gets cut or windowed every measurement. Windowing equals a product of the signal $x_s(t)$ and a window function $w(t)$ (see Fig. 2.4):

$$x_w(t) = x_s(t)\, w(t). \tag{2.22}$$

In the frequency space, windowing equals the convolution of the signal X_s and window W:

$$X_w(\omega) = X_s(\omega) * W(\omega). \tag{2.23}$$

As is shown in Fig. 2.4, we start to measure the signal at an instance of time, then use a *rectangular window function*, which equals one during the sampling period and zero else.

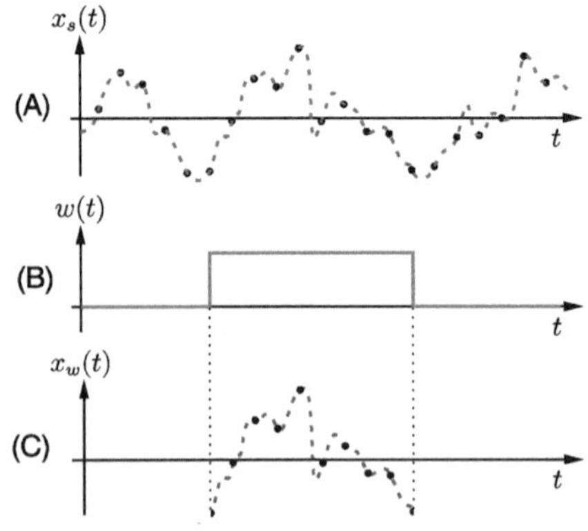

FIGURE 2.4 Windowing of a signal.

A signal with angular frequency ω is represented by Dirac functions at ω and $-\omega$. When sampling the signal with a window of length equal to the signal-period length, the indicated spectrum is true; this is because the discrete frequencies coincide with the dips in the window spectrum (Fig. 2.5). When the window length differs from the multiples of the signal period, spectral leakage occurs (Fig. 2.6). In real-world scenarios the signal consists of multiple frequencies, and a multiple period window length that satisfies all the frequencies does not exist. Therefore spectral leakage is to be expected.

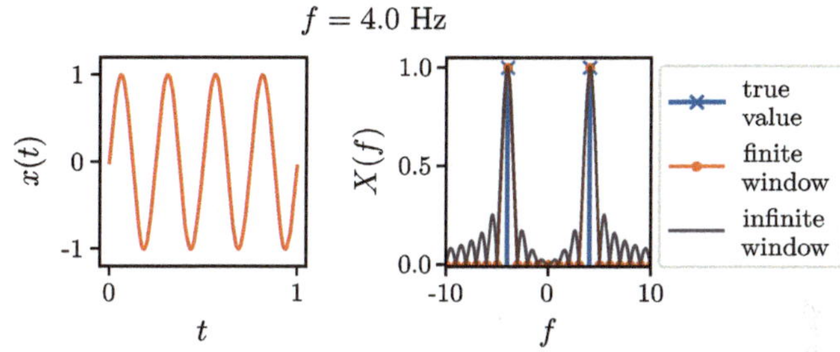

FIGURE 2.5 A rectangular window of multiple period length.

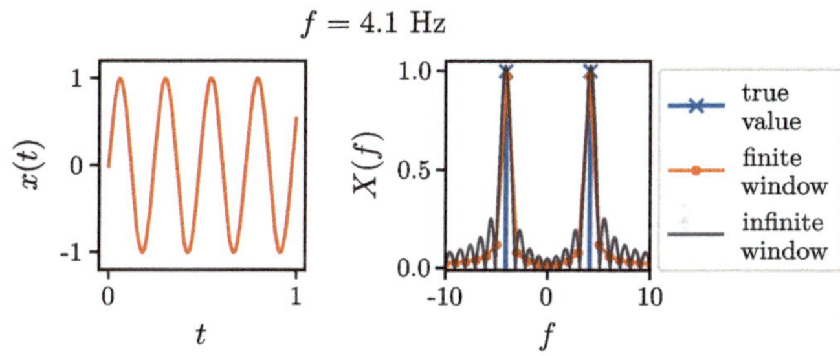

FIGURE 2.6 A rectangular window of nonmultiple period length.

To mitigate the errors in the interpretation of the signal in the frequency domain, various *window*[2] *functions* are used. Fig. 2.7 shows the time- and frequency-domain influence of some of the most frequently used window functions; for a more elaborate analysis, we refer the reader to [16].

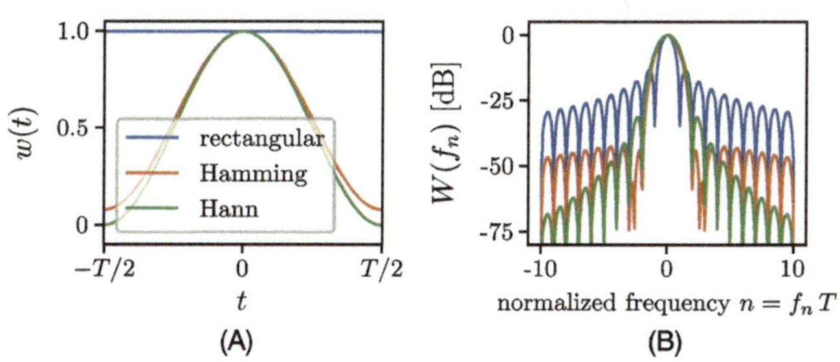

FIGURE 2.7 Some window functions in (A) time domain and (B) frequency domain.

---

2 Or *weight functions*.

Fig. 2.8 shows the influence of window length on the measured spectrum of an impulse response. Fig. 2.8A shows various length (e.g., $L = 10000$ samples) signals produced by Eq. (1.192), and Fig. 2.8B shows the spectrum obtained with a discrete Fourier transform of various length responses. The real receptance $\alpha(\omega)$ is added for comparison (1.189). By shortening the signal the information is lost, and the receptance gets erroneous. When measuring an impulse response, a good window function is an exponential window. The exponential window minimizes the end of the signal whilst adding apparent damping, which can be compensated for in the process of modal parameter identification [21].

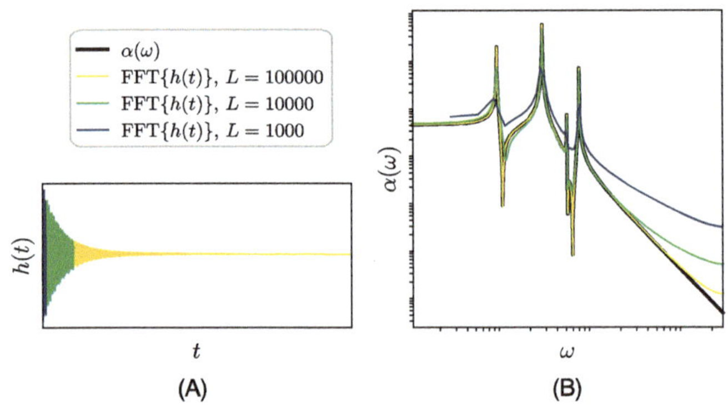

FIGURE 2.8 The influence of window length on an FRF when measuring the impulse response.

Fig. 2.9 shows the influence of the choice of the windowing function on the measured FRF in the case of measuring the response of a structure to white noise excitation. The data in the figure were simulated and do not include any measurement noise or any other error beside the windowing effect. The FRF appears to be noisy due to spectral leakage of nearby leaked frequency components,

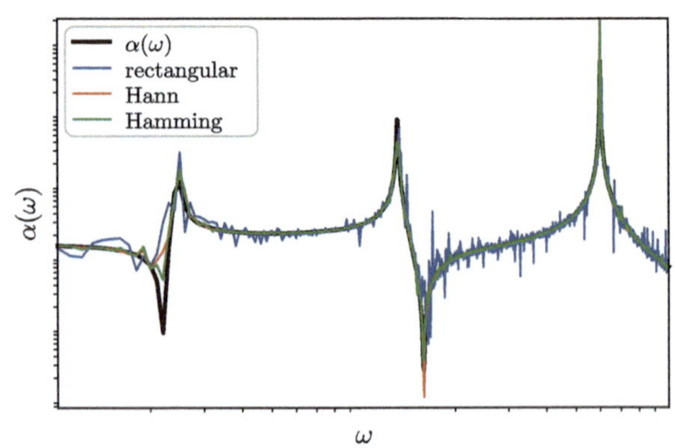

FIGURE 2.9 The effect of window choice on the FRF when measuring a response to white noise excitation.

which are of a random phase and therefore act constructively or destructively, thereby producing a ragged FRF. White noise is used when exciting structures with shakers. Due to the spectral leakage, the use of true white noise is not advisable when trying to measure the modal information; in such cases a pseudorandom (multisine) signal is preferable (see Section 3.1).

2.2 Random process

Vibration fatigue often deals with loads originating from environment, such as sea waves and currents, road roughness, and propulsion engine vibration. These loadings must be treated as random processes to be properly defined and used successfully throughout the vibration fatigue analysis. Structures that can be assumed linear respond with a broadband response when excited with broadband (random) vibrations. The response can be defined either by kinematic quantities or by stresses and strains, but always with variables of random nature. The excitation and response, as referred to in the area of vibration fatigue analysis, are in general time-dependent types of random processes that span a considerable range of the frequency content and warrant a rigorous mathematical treatment.

Random data cannot, in contrast with deterministic data, be precisely predicted, until they are obtained. The broader properties of random data can be described, but that still does not allow a reliable estimation of the future outcome. It is possible to deduce certain characteristics of random data by analyzing their segments, e.g., measuring wind velocity or surface roughness. Based on deductions from measurements, some expectations about the yet unobserved random process can be formed.

Broadband *excitation* and stress *response* are commonly defined in the frequency domain (e.g. the power spectral density, PSD, specifying the power distribution over the frequency band). Accepting the additional assumption that the amplitudes are normally distributed gives the form of the random process definition typical for vibration fatigue analysis, but also for Gaussian random processes in general. The reference texts examining this subject in-depth, recommended to the reader, are by Bendat and Piersol [15], Newland [17], and Shin and Hammond [16].

Although a random process can be analyzed in the time domain as well, there are some drawbacks for using such an approach, especially for the case of vibration fatigue analysis. The amplitude at a point in time is known with certainty given a deterministic time history of the loading process. However, in a random load, this amplitude will appear only as one possible realization of a random variable. To adequately evaluate a process, a sufficiently large number of sample measurements (also records or observations) must be taken in the time domain, which are then to be analyzed as a group (ensemble). Such a large amount of data is cumbersome to analyze, compared to the frequency-domain description of the process. Another strong argument for the frequency domain vibration fatigue is the natural expansion to structural dynamics (e.g., modal analysis): for linear systems, the random input is related to the random response using modal properties of the analyzed structure.

A random process is defined through a combination of the *amplitude probability density function* (PDF) and the *power spectral density*. The definition, in a very broad sense, corresponds to an ensemble $\{x_k(t)\}$ of sample functions (observations) $x_k(t)$, each observation k consisting of a random variable $x_k(t_i)$ at time t_i; see Fig. 2.10. As will be discussed later, the assumptions on stationarity and ergodicity significantly simplify the random data analysis. These assumptions do not fit every random process that might be relevant for vibration fatigue analysis but are necessary in making the mathematical formulations manageable and are thus regularly applied.

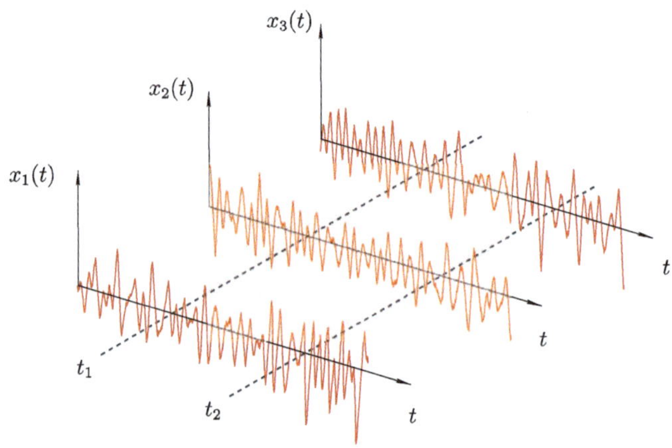

FIGURE 2.10 An ensemble of time histories (sample functions) of a random process, denoted by $\{x_k(t)\}$.

2.2.1 Gaussian distribution

Gaussian distribution is often observed in different physical phenomena; its prevalence is explained by the central limit theorem [15]. In general, a Gaussian process is such that the probability of a combination of random variables at fixed times $\{x(t_i)\}$ follows a multidimensional Gaussian distribution. Looking at only one instant in time t, the probability of the amplitude x is governed by the normal PDF

$$p(x) = \frac{1}{\sigma\sqrt{2\pi}}\,\mathrm{e}^{-\frac{(x-\mu)^2}{2\sigma^2}}, \tag{2.24}$$

shown in Fig. 2.11. Here μ is the mean value, and σ (or σ^2) is the standard deviation (or variance), respectively. The mean value μ and variance σ^2 define the shape of the curve and are often named the *first moment* and the *second central moment*, respectively, as they can be obtained by calculating the moments of a known PDF:

$$\mu = \int_{-\infty}^{+\infty} x\, p(x)\, \mathrm{d}x, \tag{2.25}$$

$$\sigma^2 = \int_{-\infty}^{+\infty} (x-\mu)^2\, p(x)\, \mathrm{d}x. \tag{2.26}$$

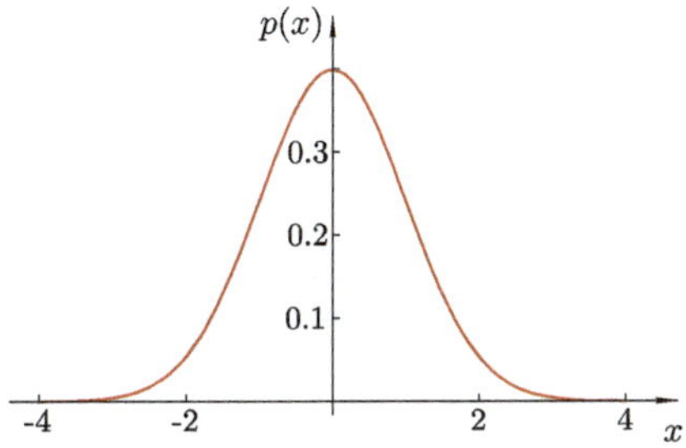

FIGURE 2.11 Univariate Gaussian probability density function with zero mean value ($\mu = 0$) and standard deviation $\sigma = 1$.

2.2.2 Moments of amplitude probability density function

Statistical quantities that are used to convey the general properties of the random processes are given as follows. In the case of two (or more) processes $\{x_k(t)\}$ and $\{y_k(t)\}$ the joint properties are sought for generality. These can represent, for instance, two components of the stress tensor. The first basic quantity of interest is the *ensemble mean value*, calculated over the index k:

$$\begin{aligned} \mu_x(t) &= E\left[x(t)\right], \\ \mu_y(t) &= E\left[y(t)\right], \end{aligned} \tag{2.27}$$

which in general varies with time t. The statistical expectation $E[x]$ signifies that the mean value of $x(t)$ was estimated by averaging over the index k (over the ensemble).

Covariance functions are defined for two processes $\{x_k(t)\}$ and $\{y_k(t)\}$ as

$$\begin{aligned} C_{xx}(t, t+\tau) &= E\left\{\left[x(t)-\mu_x(t)\right]\left[x_k(t+\tau)-\mu_x(t+\tau)\right]\right\}, \\ C_{yy}(t, t+\tau) &= E\left\{\left[y(t)-\mu_y(t)\right]\left[y_k(t+\tau)-\mu_y(t+\tau)\right]\right\}, \\ C_{xy}(t, t+\tau) &= E\left\{\left[x(t)-\mu_x(t)\right]\left[y_k(t+\tau)-\mu_y(t+\tau)\right]\right\} \end{aligned} \tag{2.28}$$

and are again dependent on time t. A special case worth noting is at $\tau = 0$:

$$\begin{aligned}
C_{xx}(t,t) &= E\left\{[x_k(t)-\mu_x(t)]^2\right\} = \sigma_x^2(t), \\
C_{yy}(t,t) &= E\left\{\left[x_y(t)-\mu_y(t)\right]^2\right\} = \sigma_y^2(t), \\
C_{xy}(t,t) &= E\left\{[x_k(t)-\mu_x(t)]\left[y_k(t)-\mu_y(t)\right]\right\} = C_{xy}(t).
\end{aligned} \tag{2.29}$$

The variances $\sigma_x^2(t)$ and $\sigma_y^2(t)$ are thus defined, and $C_{xy}(t)$ is the covariance between $\{x_k(t)\}$ and $\{y_k(t)\}$ at time t. If the analyzed processes fit a two-dimensional normal distribution, the listed quantities of Eqs. (2.27) and (2.29) are sufficient to fully describe the probability at specific points in time t, and higher-order quantities need not to be defined.

The random processes $\{x_k(t)\}$ and $\{y_k(t)\}$ are said to be *weakly stationary* when the mean values and covariance functions are time-independent. They are considered *strongly stationary* when the higher-order moments and joint moments are also time-independent. Weak and strong stationarity coincide for a Gaussian process because the probability distributions can be derived solely from mean values and covariances [17].

Autocorrelation $R_{xx}(\tau)$ and *cross-correlation* functions $R_{xy}(\tau)$ are introduced for stationary random processes, which are equal to the covariance functions when the processes have zero mean:

$$\begin{aligned}
R_{xx}(\tau) &= E\left[x_k(t)\, x_k(t+\tau)\right], \\
R_{yy}(\tau) &= E\left[y_k(t)\, y_k(t+\tau)\right], \\
R_{xy}(\tau) &= E\left[x_k(t)\, y_k(t+\tau)\right].
\end{aligned} \tag{2.30}$$

The expected value of x at specific time t can also be determined from its probability distribution $p(x)$, when such a distribution is known, according to Eqs. (2.25) and (2.26):

$$E\left[x(t)\right] = \int_{-\infty}^{+\infty} x\, p(x)\,\mathrm{d}x = \mu_x(t). \tag{2.31}$$

The variance σ_x^2 or second central moment, calculated by definition about the mean value, can also be calculated accordingly:

$$E\left[(x(t)-\mu_x)^2\right] = \int_{-\infty}^{+\infty} (x-\mu_x)^2\, p(x)\,\mathrm{d}x = \sigma_x^2(t). \tag{2.32}$$

The vibration response has typically zero mean value, and therefore we will conveniently assume that the mean of the process is zero, and consequently the variance σ_x^2 is equal to the expected square value $x^2(t)$:

$$E\left[x^2(t)\right] = \int_{-\infty}^{+\infty} x^2\, p(x)\,\mathrm{d}x = R_{xx}(0). \tag{2.33}$$

2.2.3 Ergodicity

The mean values can be derived as the *time average* instead of *ensemble average*:

$$\begin{aligned}\mu_x(k) &= \lim_{T\to\infty}\frac{1}{T}\int_0^T x_k(t)\,\mathrm{d}t,\\ \mu_y(k) &= \lim_{T\to\infty}\frac{1}{T}\int_0^T y_k(t)\,\mathrm{d}t,\end{aligned}\tag{2.34}$$

and similarly for the covariance functions:

$$\begin{aligned}C_{xx}(\tau,k) &= \lim_{T\to\infty}\frac{1}{T}\int_0^T [x_k(t)-\mu_x(k)]\,[x_k(t+\tau)-\mu_x(k)]\,\mathrm{d}t\\ &= R_{xx}(\tau,k)-\mu_x^2(k),\\ C_{yy}(\tau,k) &= \lim_{T\to\infty}\frac{1}{T}\int_0^T \left[y_k(t)-\mu_y(k)\right]\left[y_k(t+\tau)-\mu_y(k)\right]\mathrm{d}t\\ &= R_{yy}(\tau,k)-\mu_y^2(k),\\ C_{xy}(\tau,k) &= \lim_{T\to\infty}\frac{1}{T}\int_0^T [x_k(t)-\mu_x(k)]\left[y_k(t+\tau)-\mu_y(k)\right]\mathrm{d}t\\ &= R_{xy}(\tau,k)-\mu_x(k)\,\mu_y(k).\end{aligned}\tag{2.35}$$

The process is then said to be *weakly ergodic* if the *time averages* equal *ensemble averages*, independently of the chosen k:

$$\begin{aligned}\mu_x(k) &= \mu_x,\\ \mu_y(k) &= \mu_y,\\ C_{xx}(\tau,k) &= C_{xx}(\tau),\\ C_{yy}(\tau,k) &= C_{yy}(\tau),\\ C_{xy}(\tau,k) &= C_{xy}(\tau).\end{aligned}\tag{2.36}$$

If all the higher order statistical properties also exhibit this same phenomena, the process is said to be *strongly ergodic*. For the Gaussian distribution, strong and weak ergodicity are interchangeable terms, as was the case with stationarity – the first and second central moments are sufficient to uniquely describe the Gaussian distribution. In such cases, statistical properties of each separate sample function are representative for the complete ensemble. In further analysis of ergodic processes the k index can therefore be omitted, and the sample function, completely describing the properties of the random process, is denoted $x(t)$.

The ergodicity is relevant due to various reasons: it further simplifies theoretical treatment of random processes, but more importantly, it makes analysis of actual measured random data feasible. It would be relatively cumbersome and time-consuming to always analyze a large ensemble of time histories. Instead, it is usually sufficient to look at a single signal record and, relying on the ergodicity assumption, extract from it the needed statistical properties.

Throughout the text, we further assume that the process has zero mean and therefore the variance σ_x^2 can be calculated in a couple of different ways:

$$\begin{aligned}\sigma_x^2 = E\left[x^2(t)\right] &= \int_{-\infty}^{+\infty} x^2\, p(x)\,\mathrm{d}x \\ &= \lim_{T\to\infty}\frac{1}{2T}\int_{-T}^{T} x^2(t)\,\mathrm{d}t \\ &= \int_{-\infty}^{+\infty} S_{xx}(f)\,\mathrm{d}f.\end{aligned} \tag{2.37}$$

This shows that the missing property σ_x^2, which fully describes the probabilistic nature of the process, can be obtained from the power spectral density (PSD) $S_{xx}(f)$ in the frequency domain (see also (2.7)). The details of the PSD are discussed in the next section.

2.2.4 Power spectral density (PSD)

The power spectral density describes how power is distributed over the frequency content of the random process. It complements the probability density function in the definition of a specific random process. An intuitive and historically important way of obtaining a PSD was by filtering and averaging random process. Abundant processing capabilities of modern computers make it possible to obtain the PSD by means of the Fourier transform (2.2), commonly using the FFT algorithm of the DFT (2.3). The caveat is that a typical random process, as it is theoretically defined, is not periodic and extends to infinity in the time domain, meaning that the following Dirichlet condition is not fulfilled [16]:

$$\int_{-\infty}^{+\infty} |x(t)|\,\mathrm{d}t < \infty. \tag{2.38}$$

The formal theoretical definition of the PSD, employing the Fourier transform, is then based on the autocorrelation function $R_{xx}(\tau)$, which implicitly contains the frequency content of $x(t)$ while also satisfying the condition from Eq. (2.38), at least for the case of zero-mean processes. Since the autocorrelation function and the PSD are a Fourier transform pair, the following (Wiener–Khinchin) relations can be written for the random processes $x(t)$ and $y(t)$:

$$\begin{aligned}S_{xx}(f) &= \mathcal{F}\{R_{xx}(\tau)\}, \\ S_{yy}(f) &= \mathcal{F}\{R_{yy}(\tau)\}, \\ S_{xy}(f) &= \mathcal{F}\{R_{xy}(\tau)\}.\end{aligned} \tag{2.39}$$

The PSDs $S_{xx}(f)$ and $S_{yy}(f)$ (also called *autospectral densities*) are even, real-valued, and positive functions. The *cross-spectral density* $S_{xy}(f)$ is in general a complex-valued function.

In real experiments, the measured time history is of finite length, and the condition from Eq. (2.38) is fulfilled, making it possible to estimate the PSD directly on the Fourier transform of the time history ensemble:

$$\begin{aligned} X_k(f,T) &= \int_0^T x_k(t)\,\mathrm{e}^{-\mathrm{i}\,2\pi\,f\,t}\,\mathrm{d}t, \\ Y_k(f,T) &= \int_0^T y_k(t)\,\mathrm{e}^{-\mathrm{i}\,2\pi\,f\,t}\,\mathrm{d}t, \end{aligned} \tag{2.40}$$

and then multiplying:

$$S_{xy}(f,T,k) = \frac{1}{T}\,X_k^*(f,T)\,Y_k(f,T) \tag{2.41}$$

for one sample function (time history) k within the ensemble; see also Eq. (2.17). The final result is then obtained by averaging along the ensemble, giving the expected value of the PSD:

$$S_{xy}(f) = \lim_{T\to\infty} E\left[S_{xy}(f,T,k)\right]. \tag{2.42}$$

Alternatively, the PSD estimate can be evaluated on a single time history sample for the case of an ergodic process (without the index k):

$$\begin{aligned} S_{xx}(f) &= \frac{1}{T}\,X^*(f)\,X(f) \qquad \text{(PSD)}, \\ S_{xy}(f) &= \frac{1}{T}\,X^*(f)\,Y(f) \qquad \text{(CSD)}, \end{aligned} \tag{2.43}$$

where $X(f)$ and $Y(f)$ are obtained using (2.2), and Eq. (2.43) is the definition of PSD (or autospectral density) and cross-spectral density (CSD).

The autospectral density S_{xx} covers negative and positive frequencies and is therefore called the *two-sided* spectral density function. Frequently, a *one-sided* spectral density function G_{xx} is used (see also Eq. (2.16)):

$$G_{xx} = \begin{cases} 2\,S_{xx}(f), & f > 0, \\ S_{xx}(f), & f = 0, \\ 0, & f < 0. \end{cases} \tag{2.44}$$

2.2.5 Spectral moments

Due to the definition of the PSD, it follows that the area under the curve is equal to the expected square value $E\left[x^2\right]$ of the process (2.37):

$$m_0 = E\left[x^2(t)\right] = \int_{-\infty}^{+\infty} S_{xx}(\omega)\,\mathrm{d}\omega, \tag{2.45}$$

where m_0 is the zeroth spectral moment equal to the *variance* σ_x^2 of the zero-mean process (2.7). Further, higher even-order moments correspond to the variances of the time derivatives of the original random process (see also (2.13)):

$$m_2 = E\left[\dot{x}^2(t)\right] = \int_{-\infty}^{+\infty} \omega^2\, S_{xx}(\omega)\,\mathrm{d}\omega. \tag{2.46}$$

A general expression for the ith spectral moment m_i can be written in the form

$$m_i = \int_{-\infty}^{+\infty} \omega^i\, S_{xx}(\omega)\mathrm{d}\omega = \int_{0}^{+\infty} \omega^i\, G_{xx}(\omega)\,\mathrm{d}\omega, \tag{2.47}$$

where G_{xx} is the one-sided PSD (2.44). Point out that if the PSD is given in Hz (i.e., $G_{xx}(f)$), then the 2π needs not to be forgotten in the spectral moment:

$$m_i = \int_{0}^{+\infty} (2\pi\, f)^i\, G_{xx}(f)\,\mathrm{d}f. \tag{2.48}$$

2.2.6 Bandwidth of a random process

The distribution of power over the frequency axis is uniquely important in fatigue damage evaluation due to the significance of the damaging cycles and the way in which they arise in the time history of the process. This is most obvious when comparing the time-domain features of a decidedly narrowband process and a broadband process, as seen in Fig. 2.12. The narrowband process has power concentrated in a very narrow frequency band, which manifests in the time-domain approximately in the form of an amplitude-modulated harmonic function. *Cycle amplitudes* are easily defined for this type of signal in both time and frequency domains, which will be explained in detail in Sec. 4.3.

Broadband process makes it harder to identify specific cycles and their amplitudes in the time domain and requires special spectral moment methods to derive an estimate of the distribution of cycle amplitudes in the frequency domain. Estimators are commonly used to describe the narrowband and the broadband process, and all others in-between, by assigning them a scalar number between zero and one.

Later, we will use the *narrowband/broadband estimator* α_i:

$$\alpha_i = \frac{m_i}{\sqrt{m_0\, m_{2i}}}, \tag{2.49}$$

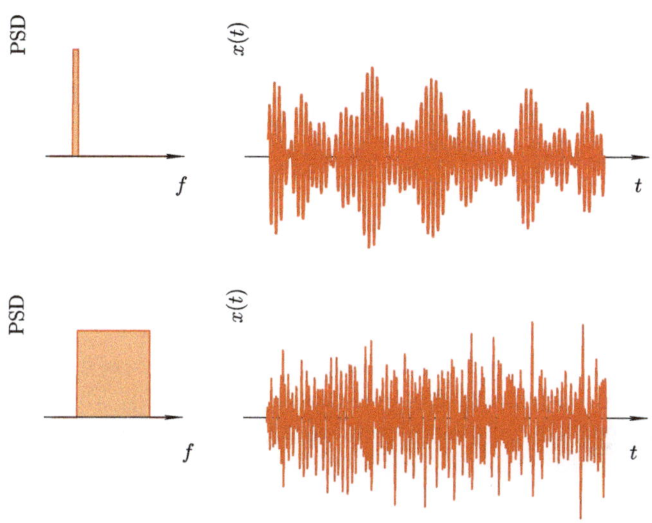

FIGURE 2.12 A side-by-side comparison of a narrow- and broadband signal in the frequency and time domains.

where i can be any positive real number, but it is usually chosen as 1, 2, or 0.75. Wirsching and Light [22] used the α_2 estimator to describe narrowband signals (α_2 is closer to 1) or broadband signals (α_2 closer to 0):

$$\epsilon = \sqrt{1 - \alpha_2^2}. \tag{2.50}$$

The *Vanmarcke bandwidth parameter* ϵ_V is another established estimator defined using α_1 [23]:

$$\epsilon_V = \sqrt{1 - \alpha_1^2}. \tag{2.51}$$

When derivatives of the process are deemed important, the parameter β_i is used, which can be thought of as α_i calculated for derivative $\dot{x}(t)$ and is defined as

$$\beta_i = \frac{m_{i+2}}{\sqrt{m_2 m_{2i+2}}}. \tag{2.52}$$

2.2.7 Peak and level-crossing frequency

Fatigue analysis is based largely on the local extreme values of a signal and the implied damaging cycles, as observed in the time domain. Those phenomena are not present in the frequency domain, and therefore special estimators are devised to describe them in a probabilistic way. This is done based on the mathematical description of the Gaussian probability and statistical properties that can be deduced from the PSD, such as variances of the random process and its derivatives:

$$m_0 = \sigma_x^2, \qquad m_2 = \sigma_{\dot{x}}^2, \qquad m_4 = \sigma_{\ddot{x}}^2. \tag{2.53}$$

If, for example, σ_x^2 is the variance of the displacement, then $\sigma_{\dot{x}}^2$ and $\sigma_{\ddot{x}}^2$ are the variances of velocity and acceleration, respectively.

Frequency of positive-slope level-crossings

The first step in properly characterizing the random process for the purpose of fatigue analysis is defining the level-crossing probability. The expression for the probability of $x(t)$ crossing level a is studied on the short time interval $\mathrm{d}t$, shown in Fig. 2.13.

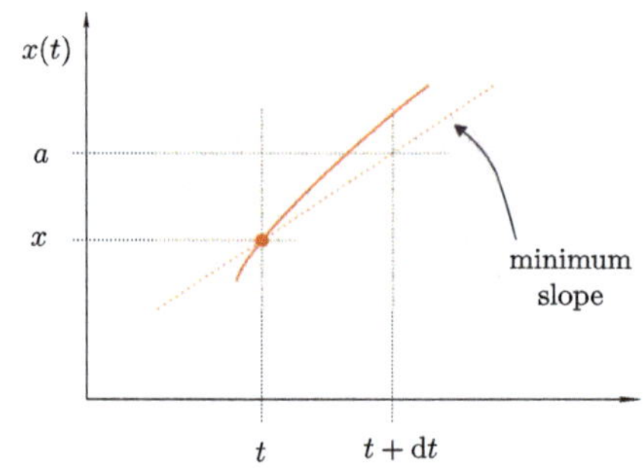

FIGURE 2.13 A short time interval $\mathrm{d}t$ of the sample function $x(t)$ in the time domain.

For a small enough $\mathrm{d}t$, the sample function will only be able to reach the level a in the positive direction when $x < a$, and the slope is steeper than the minimum required for it to reach a within $\mathrm{d}t$. These two conditions are written in the form of

$$\begin{aligned} x &< a, \\ \dot{x} &> \frac{a - x}{\mathrm{d}t}. \end{aligned} \tag{2.54}$$

Assuming that the joint probability density function $p(x, \dot{x})$ is known, the area bound by the proposed conditions can be defined. The probability of the level a crossing occurring with a positive slope within the $\mathrm{d}t$ is equal to the slice of volume underneath the surface $p(x, \dot{x})$:

$$\Pr\{\text{level } a \text{ crossing}\} = \int_0^{\infty} \left(\int_{a-\dot{x}\,\mathrm{d}t}^{a} p(x, \dot{x})\,\mathrm{d}x \right) \mathrm{d}\dot{x}. \tag{2.55}$$

For the short interval $\mathrm{d}t \to 0$, we can assume that $x = a$ and therefore $p(x, \dot{x})$ is constant with regard to integration over x, and it follows that

$$\Pr\{\text{level } a \text{ crossing}\} = \mathrm{d}t \int_0^\infty p(a, \dot{x})\, \dot{x}\, \mathrm{d}\dot{x}. \tag{2.56}$$

If the $\mathrm{d}t$ is indeed very small, then the crossing will either occur once or not at all. The probability of level crossing is equal to the expected frequency of level crossings, ν_a^+:

$$\nu_a^+\, \mathrm{d}t = \mathrm{d}t \int_0^\infty p(a, \dot{x})\, \dot{x}\, \mathrm{d}\dot{x}. \tag{2.57}$$

It follows from the definitions of the autocorrelation and cross-correlation functions that for a zero-mean stationary random data, the covariance $\sigma_{x\dot{x}}$ for the process $x(t)$ and its derivative $\dot{x}(t)$ equals 0:

$$R_{x\dot{x}}(0) = E\left[x(t)\, \dot{x}(t)\right] = 0. \tag{2.58}$$

This fact simplifies the joint probability density function $p(a, \dot{x}) = p(a)\, p(\dot{x})$. Using the normal probability density function (2.24) with Eq. (2.57) for a zero-mean process, we have

$$\nu_a^+ = \frac{1}{\sqrt{2\pi}\,\sigma_x} \exp\left(\frac{-a^2}{2\,\sigma_{x^2}}\right) \int_0^\infty \frac{1}{\sqrt{2\pi}\,\sigma_{\dot{x}}} \exp\left(\frac{-\dot{x}^2}{2\,\sigma_{\dot{x}^2}}\right) \dot{x}\, \mathrm{d}\dot{x}, \tag{2.59}$$

where the integral is equal to $\sigma_{\dot{x}}/\sqrt{2\pi}$. The final expression for the *expected mean frequency of positive-slope level-crossing*, written using spectral moments (2.47), is

$$\nu_a^+ = \frac{1}{2\pi} \sqrt{\frac{m_2}{m_0}}\, \mathrm{e}^{-\frac{a^2}{2\,m_0}}. \tag{2.60}$$

If we are interested in the *frequency of positive- and negative-slope level-crossing*, then

$$\nu_a^\pm = 2\, \nu_a^+ = \frac{1}{\pi} \sqrt{\frac{m_2}{m_0}}\, \mathrm{e}^{-\frac{a^2}{2\,m_0}}. \tag{2.61}$$

A special form of this result is for the *frequency of positive slope zero crossing*, which is obtained by setting $a = 0$:

$$\nu_0^+ = \frac{1}{2\pi} \sqrt{\frac{m_2}{m_0}}. \tag{2.62}$$

Peak probability

The expected frequencies ν_a^+ (2.60) and ν_0^+ (2.62) are with regard to the instantaneous amplitudes. To finally evaluate fatigue damage, the cycle amplitudes should be determined. As a first step in this direction, we define a PDF $p_{\mathrm{p}}(a)$, which describes the probability of occurrence of a peak with amplitude a or larger:

$$\Pr\{\text{peak} > a\} = \int_a^\infty p_{\mathrm{p}}(a)\,\mathrm{d}a. \tag{2.63}$$

For the case of a zero-mean narrowband process, each positive slope zero crossing accounts for one cycle. This is evident by looking at the time history, where it resembles a kind of an amplitude-modulated harmonic function (see Fig. 2.12).

If ν_0^+ is the expected peak frequency and ν_a^+ is the expected frequency of amplitude being larger than a, then the probability that the peak is larger than a is

$$\int_a^\infty p_{\mathrm{p}}(\tilde{a})\,\mathrm{d}\tilde{a} = \frac{\nu_a^+}{\nu_0^+}. \tag{2.64}$$

Using the expressions for ν_a^+ (2.60) and ν_0^+ (2.62) and then differentiating Eq. (2.64) with respect to a, we define the PDF as

$$p_{\mathrm{p}}(a) = \frac{a}{m_0}\,\mathrm{e}^{-\frac{a^2}{2\,m_0}}, \tag{2.65}$$

which is the Rayleigh PDF and only valid for those narrowband processes that satisfy the assumption of exactly one peak occurring for every positive-slope zero-crossing. Fig. 2.14 shows (2.65): small and large peaks have relatively small density, whereas the highest density have the amplitudes close to $\sigma_x = \sqrt{m_0}$.

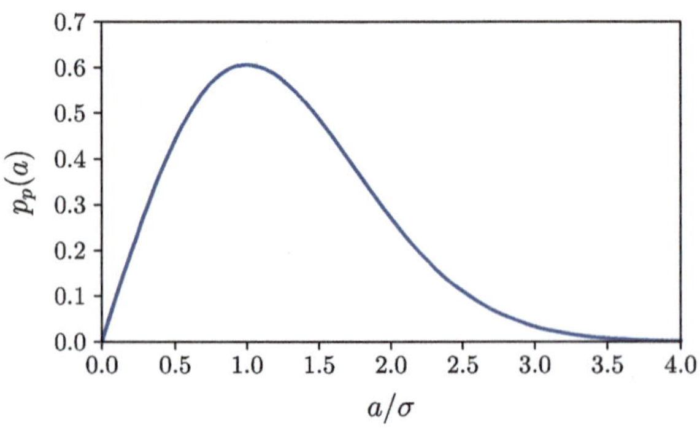

FIGURE 2.14 Probability density distribution of peak amplitudes for a narrowband random processes.

For a more general expression also valid for a broadband process, the conditions for positive extrema must be met: x is a peak when $\dot{x} = 0$ and $\ddot{x} < 0$. This holds at the instant of the negative-slope zero-crossing, observed in the first derivative of the process, giving the expected peak frequency $\nu_\mathrm{p} = \nu^-_{\dot{x}=0}$. We make the assumption that the number of positive slope zero-crossings is equal to the number of negative zero-crossings, $\nu^-_{\dot{x}=0} = \nu^+_{\dot{x}=0}$. The *expected peak frequency* is then given by writing Eq. (2.62) for the derivative $\dot{x}(t)$ using higher spectral moments:

$$\nu_\mathrm{p} = \frac{1}{2\pi}\sqrt{\frac{m_4}{m_2}}, \tag{2.66}$$

considering $\sigma^2_{\dot{x}} = m_2$ and $\sigma^2_{\ddot{x}} = m_4$.

Assuming that Eq. (2.65) is valid for a narrowband process, Rice [18] provides a PDF of the peak amplitudes $p_\mathrm{p}(a)$ for a broadband process:

$$p_\mathrm{p}(a) = \frac{\sqrt{1-\alpha_2^2}}{\sqrt{2\pi}\,\sigma_x}\,\mathrm{e}^{-\frac{a^2}{2\sigma_x^2\left(1-\alpha_2^2\right)}} + \frac{\alpha_2\, a}{\sigma_x^2}\,\mathrm{e}^{-\frac{a^2}{2\sigma_x^2}}\,\Phi\left(\frac{\alpha_2\, a}{\sigma_x\sqrt{1-\alpha_2^2}}\right), \tag{2.67}$$

where α_2 is the narrowband/broadband estimator (2.49) of the process, and Φ is the standard normal cumulative distribution function ($\mu = 0$ and $\sigma_x = 1$):

$$\Phi(z) = \frac{1}{\sqrt{2\pi}}\int_{-\infty}^{z} \mathrm{e}^{-\frac{t^2}{2}}\,\mathrm{d}t. \tag{2.68}$$

The function $p_\mathrm{p}(a)$ from Eq. (2.67) is shown in Fig. 2.15 and gives a great deal of information about the features of the random process in the time domain based on the frequency-domain data. It is clearly visible that for a narrowband processes ($\alpha_2 = 1$), the majority of peaks will appear above the process mean (similarly to Fig. 2.14). However, for a broadband process ($\alpha_2 = 0$), a larger portion of peaks will have negative amplitudes.

However, the peak probability density does not yet make it possible to derive the rainflow amplitude distribution. Rather, it is one part of a framework in which, finally, the rainflow amplitude probability can be estimated using spectral moment methods introduced in Chapter 4. Those are usually empirical models based on simulations and, in large part, on the estimators of the broadband random process derived here.

2.2.8 Linear time-invariant system response to random input

A linear time-invariant system (1.58) is defined with the superposition principle (1.60) and the time-invariance (1.61) property. The linear time-invariant system is characterized by the impulse response function $h(t)$ (1.79); with regard to

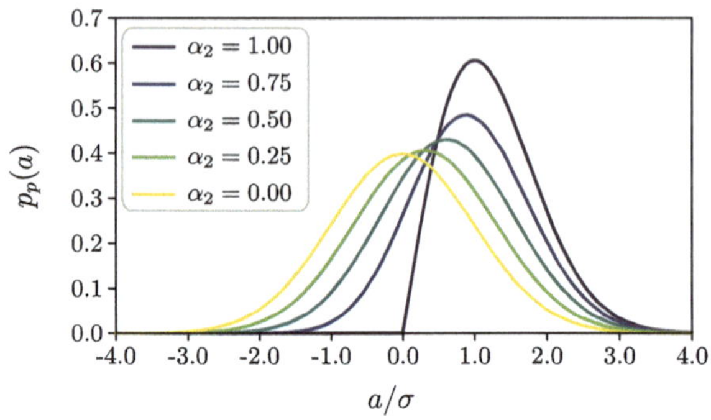

FIGURE 2.15 Probability density distribution of peak amplitudes for random processes characterized by different values of α_2.

the excitation/input $f(t)$, the response $x(t)$ is defined as the convolution (1.85):

$$x(t) = f(t) * h(t) = \int_{-\infty}^{t} f(t)\, h(t-\tau)\, \mathrm{d}\tau = \int_{0}^{+\infty} h(\tau)\, f(t-\tau)\, \mathrm{d}\tau, \quad (2.69)$$

where we have assumed that the excitation $f(t)$ is *stationary*.

Response mean value

If μ_f is the mean value of the excitation, then the mean value of the response μ_x is obtained as (2.27)

$$\mu_x = E\left[\int_0^{+\infty} h(\tau)\, f(t-\tau)\, \mathrm{d}\tau\right]. \quad (2.70)$$

The expectation operator $E[\cdot]$ is linear, and therefore the Eq. (2.70) simplifies to

$$\mu_x = \int_0^{+\infty} h(\tau)\, E\left[f(t-\tau)\right]\, \mathrm{d}\tau = \int_0^{+\infty} h(\tau)\, \mu_f\, \mathrm{d}\tau \quad (2.71)$$

and, finally,

$$\mu_x = \mu_f \int_0^{+\infty} h(\tau)\, \mathrm{d}\tau. \quad (2.72)$$

From Eq. (2.72) it follows that for a zero-mean excitation, the response has zero-mean regardless of $h(t)$.

Autocorrelation function

If the excitation is characterized by the autocorrelation function R_{ff}, what is the autocorrelation of the response R_{xx}? With regard to Eq. (2.30) and the linearity of expectation operation, we have

$$
\begin{aligned}
R_{xx}(\tau) &= E\left[x(t)\,x(t+\tau)\right] \\
&= E\left[\int_0^{+\infty}\int_0^{+\infty} h(\tau_1)\, f(t-\tau_1)\, h(\tau_2)\; f(t+\tau-\tau_2)\,\mathrm{d}\tau_1\,\mathrm{d}\tau_2\right] \\
&= \int_0^{+\infty}\int_0^{+\infty} h(\tau_1)\, h(\tau_2)\, E\left[f(t-\tau_1)\, f(t+\tau-\tau_2)\right]\mathrm{d}\tau_1\,\mathrm{d}\tau_2 \\
&= \int_0^{+\infty}\int_0^{+\infty} h(\tau_1)\, h(\tau_2)\, R_{ff}(\tau+\tau_1-\tau_2)\,\mathrm{d}\tau_1\,\mathrm{d}\tau_2.
\end{aligned}
\tag{2.73}
$$

Eq. (2.73) is difficult to evaluate in time-domain, and therefore we further consider the frequency-domain equivalent, the autospectral density function S_{xx}.

Autospectral density function

The autospectral density function of the response S_{xx} is defined as (2.39)

$$
S_{xx}(f) = \int_{-\infty}^{+\infty} R_{xx}(\tau)\,\mathrm{e}^{-\mathrm{i}\,2\pi\, f\,\tau}\mathrm{d}\tau. \tag{2.74}
$$

Using Eq. (2.73), we have

$$
\begin{aligned}
S_{xx}(f) = &\int_0^{+\infty} h(\tau_1)\,\mathrm{e}^{\mathrm{i}\,2\pi\, f\,\tau_1}\mathrm{d}\tau_1 \int_0^{+\infty} h(\tau_2)\,\mathrm{e}^{-\mathrm{i}\,2\pi\, f\,\tau_2}\mathrm{d}\tau_2 \\
&\times \int_{-\infty}^{+\infty} R_{ff}(u)\,\mathrm{e}^{-\mathrm{i}\,2\pi\, f\,u}\mathrm{d}u,
\end{aligned}
\tag{2.75}
$$

where

$$
\mathrm{e}^{-\mathrm{i}\,2\pi\, f\,\tau} = \mathrm{e}^{\mathrm{i}\,2\pi\, f\,\tau_1}\,\mathrm{e}^{-\mathrm{i}\,2\pi\, f\,\tau_2}\,\mathrm{e}^{-\mathrm{i}\,2\pi\, f\,(\tau+\tau_1-\tau_2)} \tag{2.76}
$$

and $u = \tau + \tau_1 - \tau_2$. As $\alpha(f) = \mathcal{F}\{h(t)\}$ (1.90), Eq. (2.75) can be rewritten as

$$
S_{xx}(f) = \alpha^*(f)\,\alpha(f)\,S_{ff}(f) = |\alpha(f)|^2\,S_{ff}(f), \tag{2.77}
$$

where $S_{ff}(f)$ is the real-valued autospectral density of the excitation (no phase information is contained).

Cross-spectral density function

Similar to (2.77), the cross-spectral density function can be obtained [16]:

$$
S_{fx}(f) = \alpha(f)\,S_{ff}(f). \tag{2.78}
$$

Whereas the autospectral density (2.77) does not contain the phase information, the cross-spectral density (2.78) does contain the phase information and is typically used to determine the frequency response function $\alpha(f) = S_{fx}/S_{ff}$.

Ordinary coherence function

The *ordinary coherence function* is defined as

$$\gamma_{fx}^2 = \frac{|S_{fx}(f)|^2}{S_{ff}(f)\, S_{xx}(f)} \tag{2.79}$$

and measures the degree of linear association between the excitation $f(t)$ and response $x(t)$. If the excitation and response are linearly related, then the coherence equals 1. If the excitation and response are not related, then the coherence equals 0. The coherence can be smaller than 1 if: the system exhibits nonlinearities, there are other sources of excitation besides $f(t)$, or the measurement of $f(t)$ and $x(t)$ is contaminated with noise.

Chapter 3

Experimental modal analysis

In Chapter 1, we discussed structural dynamics of SDOF and MDOF systems, and in Table 1.3, we discussed the spatial, modal, and response models. If the geometric and material properties of the structure are known, then numerical modal analysis using finite element methods (FEMs) results in the modal model. If we start from the response model, which is usually defined by experimental measurements, then the modal model can be obtained by *experimental modal analysis* or identification.

In Chapter 2, we discussed signal processing and gave the required theoretical background for characterization of experimental measurements. Here this theoretical background is extended to the experimental modal analysis, which deals with estimation of modal parameters from experimental measurements. To determine the modal parameters, the excitation $\mathbf{f}(t)$ and response $\mathbf{x}(t)$ have to be measured (see Fig. 1.13), and the modal parameters are then identified by fitting the data to a physical modal model.
Modal parameters can be identified even if the excitation force is not known. Modal parameter identification based only on measurements is called *operational modal analysis* and is more commonly used for larger structures, which are too large to be excited using hammers and/or shakers.

For further reading, we advise the works of Maia et al. [4], Ewins [24], Fu and He [25], McConnell and Varoto [20], and Peeters et al. [26].

3.1 Types of excitation

The dynamic response of structures can span over a wide range of frequencies. To measure the whole spectral response, all the frequencies have to be excited. There are multiple methods of exciting structures, each with their advantages and disadvantages.

3.1.1 Impact excitation

Impact excitation is done with modal hammers. A *modal hammer* is a hammer that measures the impact force with a force transducer placed between the hammer mass and the hammer tip. The hammer tip is preferably as light as possible to produce a correct force measurement. Different hammer tips can be used to produce impacts of different durations. The hardest tip will produce the shortest impact, and the softer tips produce impacts of longer durations. Ideally, the

Vibration Fatigue by Spectral Methods. https://doi.org/10.1016/B978-0-12-822190-7.00010-4

impact should be as short as possible, that is, a Dirac delta function $\delta(t)$ (1.76), which is an impulse with an integral of one and equals zero everywhere except $t = 0$. The Fourier transform of the Dirac delta function $\delta(t)$ is constant over the entire spectrum $\omega = (-\infty, \infty)$: $\mathcal{F}\{\delta(x)\} = 1$; see (1.87) for details. Therefore an infinitely short impact of the form of the Dirac delta function $\delta(x)$ theoretically excites the entire frequency spectrum.

A real impact is never infinitely short. Assuming a half sine signal, the effect of the impact length can be studied; see Fig. 3.1. As indicated in Fig. 3.1, impacts with longer durations excite primarily lower frequencies. Often, a softer hammer tip with a longer impact is preferable when only lower frequencies have to be excited. Softer hammer tips also produce smaller stresses on the impacted surface due to the lower peak force.

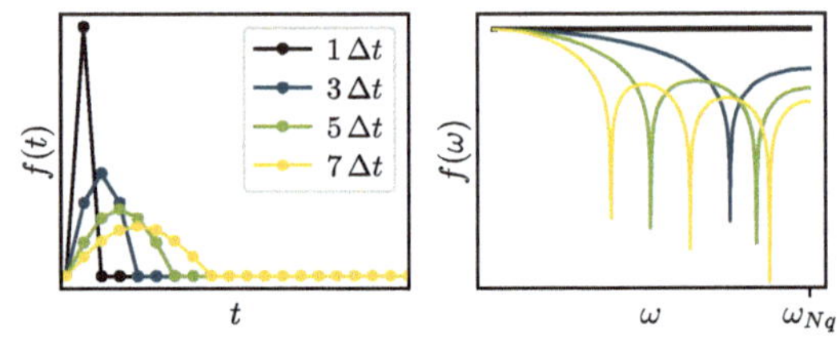

FIGURE 3.1 The effect of impact duration on the excited frequencies.

Measurements using the modal hammer can be simple and fast, because the hammer is not attached on the structure and can be moved to excite different locations at sequential measurements (also called the *roving hammer* approach). By moving the excitation point and measuring the response at a fixed point a row of the FRF matrix (1.175) can be obtained.

The less favorable feature of the impact excitation is that the structure can only be excited with a limited amount of energy before the impact force is too great to cause damage. The response following the impact is decayed, meaning that the response is the greatest after the impact and then progressively smaller, taking up a large dynamic range of the sensors used to measure it. Further, the impact excites the system in the direction perpendicular to the surface of impact only.

3.1.2 Shaker excitation

Similarly to the modal hammer, in shaker excitation the force is also measured. The shaker is usually physically attached to the structure via a stinger, a thin metal rod, to limit the cross-axial and moment excitation and to excite at discrete point only. Structures can also be attached to the shaker base directly, simulating the vibrations coming to the structure via its fixtures. Although single axis shakers are most common, multi-degree-of-freedom shakers are gaining on popularity in the recent times. Multi-degree-of-freedom shakers can excite up to all six degrees-of-freedom (3× force, 3× moment).

The benefit of using a shaker to excite a structure is that the excitation force can be a tailored signal $r(t)$. Due to the shaker's own transfer function, the *excitation signal* $r(t)$ generally differs somewhat from the excitation force $f(t)$ unless feedback loops are used to control the excitation. Using a force feedback, the excitation force will match the excitation signal $r(t)$, and using a base acceleration feedback measured with an accelerometer attached to the base of the shaker, the base excitation $\ddot{x}_{\text{base}}(t)$ will match the excitation signal $r(t)$. The first type of excitation is known as *force excitation*, and the second as the *base excitation*; see Sections 1.1.5 and 1.2.4 for theoretical details.

The control over the excitation force allows for excitation of discrete bands of frequencies. The greater control over the excitation is useful when dealing with a nonlinear response, where the excitation history influences the dynamic response. Also, compared to the modal hammer, shakers can introduce more energy into the system over a longer period of time, making shakers a better choice for larger structures and structures with a high damping.

Some of the most common excitation signals [27] are as follows:

- *Pure random* is a continuous signal with normal (Gaussian) distribution; see Section 2.2.1. The spectrum of such a signal contains all frequencies, and therefore it is a *white noise* signal. If only a limited spectral band is to be excited, then the signal should be filtered with band-pass filters. Fig. 3.2A shows the temporal and spectral plots for a pure random signal filtered with a first-order band-pass filter (a combination of low- and high-pass filters).
  The random nature of the signal means that the signal is not periodic and causes spectral leakage. When measuring the response function using a pure random excitation signal, a larger number of averages should be used along with window functions; see Section 2.1.5.
  When it comes to measuring the frequency response functions, other excitation signals (introduced bellow) are preferred.
- *Burst-random* is a pure random signal limited to a temporal burst [28]. The excitation before and after the burst should equal zero. After the burst the structure response is allowed to decay while the measurement is still running. Allowing for a decay in response can completely reduce the windowing effect that is otherwise present with random signals.
- *Pseudorandom* (also known as *Multisine*) excitation is obtained by assigning a desired amplitude spectrum $R(\omega)$ and randomly varying its phase spectrum by multiplying the amplitude spectrum with a complex exponential:

$$R(\omega) = R(\omega)\, \mathrm{e}^{\mathrm{i}\, \varphi_{\text{rand}}(\omega)}, \tag{3.1}$$

  where $\varphi_{\text{rand}}(\omega)$ is a random phase between $-\pi$ and π [28]. The excitation signal $r(t)$ is then obtained as the inverse Fourier transform of the generated spectrum $R(\omega)$. This produces a periodic signal composed of multiple harmonics with periods that are multiples of the window length. Because the signal is periodic, no windowing effects are observed when the excitation sig-

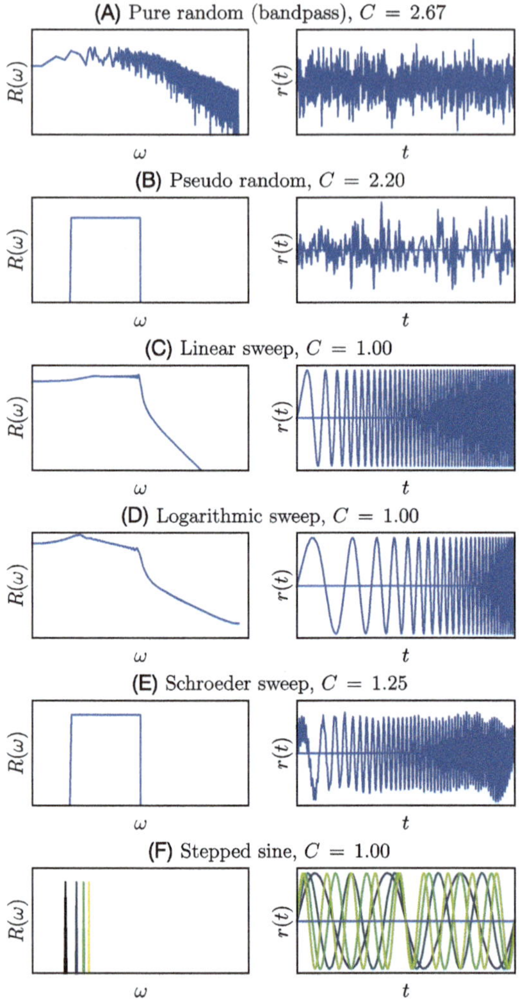

FIGURE 3.2 Shaker excitation signals.

nal window matches the measurement window; see Fig. 3.2B. Measurements should be performed only after the response becomes steady-state.

It should be noted that if the phase were left to be constant (zero), then an impulse (Dirac delta function) would be generated. By randomly varying the phase the signal expands over the entire window, and the peak force decreases.

- A *Sine sweep* excitation takes a form of a sine function with varying frequency over time (a *chirp*). An example of such an excitation are motor spin-ups.

Assume a signal $\sin\varphi(t)$, where the phase derivative takes the form

$$\varphi'(t) = 2\pi\, f(t). \tag{3.2}$$

The temporal phase $\varphi(t)$ is obtained by integrating (3.2) over time:

$$\varphi(t) = \varphi_0 + 2\pi \int_0^t f(\tau)\,\mathrm{d}\tau. \tag{3.3}$$

The constant phase φ_0 is not relevant and will be left out. By assigning a linearly increasing frequency $f(t) = f_0 + k\,t$ the integral (3.3) equals

$$\varphi(t) = 2\pi \int_0^t (f_0 + k\,\tau)\,\mathrm{d}\tau, \tag{3.4}$$

and the temporal phase is

$$\varphi(t) = 2\pi\left(f_\mathrm{s}\,t + \frac{k}{2}t^2\right), \tag{3.5}$$

where f_s is the start frequency, and k is the rate of frequency change. If a sweep from a start frequency f_s up to f_e is desired in period T, then the rate equals

$$k = \frac{f_\mathrm{e} - f_\mathrm{s}}{T}. \tag{3.6}$$

By inserting the temporal phase (3.5) into a sine function we obtain a *linear sine-sweep*:

$$r(t) = \sin\left(2\pi\left(f_\mathrm{s}\,t + \frac{f_\mathrm{e} - f_\mathrm{s}}{2T}t^2\right)\right). \tag{3.7}$$

If a different sweep shape is desired, then a different temporal frequency function should be determined, and the derivation process repeated. Along with a linear sweep, another commonly used sweep shape is the *logarithmic sweep* (sometimes referred to as *exponential sweep*). The temporal frequency for the logarithmic sweep case takes the form of $f(t) = f_0 \exp(k\,t)$. The temporal phase obtained by integrating over time is

$$\varphi(t) = \sin\left(\frac{2\pi\, f_0}{k}\left(\mathrm{e}^{k\,t} - 1\right)\right), \tag{3.8}$$

and for a sweep from f_z to f_k, the logarithmic rate k should be

$$k = \frac{1}{T}\ln\left(\frac{f_\mathrm{e}}{f_\mathrm{s}}\right). \tag{3.9}$$

Because the temporal frequency in a sweep is known (Fig. 3.3), the frequency response function can, to a limited extent, be observed from the envelop of the response $x(t)$.

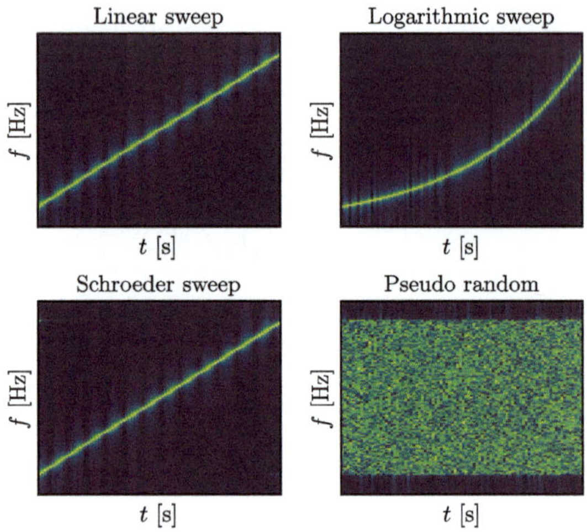

FIGURE 3.3 Spectrograms for different excitation signals.

- A *Schroeder sweep* is similar to conventional chirps, but it is derived from the frequency domain in the same way as the pseudorandom (multisine) signal. As with the pseudorandom signal, the amplitude spectrum $R(\omega)$ is assigned, the phase is varied, but unlike the pseudorandom, the phase is not random; it is determined as the realization that produces the minimal peak-to-peak value [29].

  The phase spectrum for discrete frequencies resulting from a finite window $\varphi(\omega) = (\varphi_0, \varphi_1, \varphi_2, \dots)$ is determined according to [29]

$$\varphi_{n+1} = \varphi_n - 2\pi\, P_n, \tag{3.10}$$

where P_n is the sum of the relative powers p_k of the spectrum up to the discrete frequency n:

$$P_n = \sum_{k-1}^{n} p_k. \tag{3.11}$$

A Schroeder sweep is a signal with a small peak-to-peak value and large energy. This is indicated in the *crest factor*, which is the ratio of the peak-to-peak r_{pp} and the effective value r_{rms}:

$$C = \frac{|r_{\text{pp}}|}{r_{\text{rms}}}, \tag{3.12}$$

where root mean square value *RMS* is defined as

$$r_{\text{rms}} = \sqrt{\frac{1}{T}\int_0^T r^2(t)\,\mathrm{d}t}. \tag{3.13}$$

A higher crest-factor means that the signal better covers the dynamic range of the excitation, meaning that more energy can be introduced to the structure via the shaker before clipping is reached.
As can be seen in Figs. 3.2 and 3.3, the Schroeder sweep is very similar to a chirp. In comparison with the sweep sine, the Schroeder sweep is by nature periodic and comprised of a sharply defined amplitude band, causing it to have a slightly higher crest factor than the sweep sine, but still much better than the crest factor of a pseudorandom or pure random signal.
The signal in Fig. 3.2E is produced from a single frequency band; however, any amplitude spectrum can be assigned to produce a variation of the signal.

- Another form of excitation is using *stepped sines*. A stepped sine excitation is done by exciting a structure using a sine signal of a set frequency and then repeating the measurement for different excitation frequencies. A single measurement represents a point in the spectrum. The measurements are performed at the steady-state conditions. Because the signal is periodic, no windowing occurs (if excitation is synchronized with the response measurement). Fig. 3.2F shows a variety of sines representing sequential measurements.
  Stepped sine excitation is useful when only selected frequencies should be measured (e.g., ESPI, Electronic speckle pattern interferometry [30]), or when the maximal relative power p_k is desired; otherwise, stepped sine excitation is avoided due to long measurements.

3.2 Frequency response function estimators

Frequency response functions (FRF) have been discussed theoretically; see Eq. (1.35) for different forms of the FRF. In this section, we discuss the estimation of FRFs from experimentally measured data. Initially, we discuss the single input (excitation) and the single output (response), and then we generalize the methods for the multiple-input multiple-output case.

3.2.1 Single-input single-output system

The dynamics of a linear, time-invariant, *single-input single-output* (SISO) system can be represented with a block diagram in Fig. 3.4, were $f(t)$ is the force excitation, $h(t)$ is the impulse response function (1.79), and $x(t)$ is the response; see Section 1.75; $F(\omega)$, $\alpha(\omega)$, and $X(\omega)$ are their frequency domain pairs.[1]

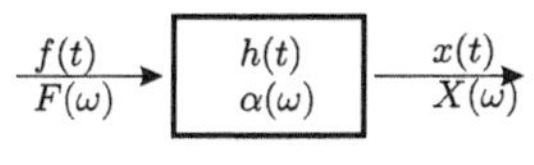

FIGURE 3.4 A block diagram for a SISO system.

The FRF for a random process is defined by $\alpha(\omega) = S_{fx}/S_{ff}$, where S_{ff} and S_{fx} are the auto- and cross-spectral density functions (2.78).

[1] A more general definition uses $H(\omega)$ as the frequency response function instead of $\alpha(\omega)$.

Fig. 3.5 shows a SISO system where $u(t)$ and $v(t)$ are the real input (excitation) and the real output (response) to the system, respectively; $m(t)$ presents the measured noise at the input $f(t)$, and $n(t)$ presents the measured noise at the output $x(t)$. Capitalized letters are the corresponding frequency domain quantities. The measurement noise is always present and distorts the true state. Noise is particularly problematic when measuring strain with strain gauges or by deriving the displacement shapes to produce strain shapes. Fig. 3.5 shows that in reality it is not possible to measure the real input to the system $u(t)$, and it is also not possible to measure the real output of the system $v(t)$. Instead, we measure the input with noise,

$$f(t) = u(t) + m(t), \tag{3.14}$$

and the response with noise,

$$x(t) = v(t) + n(t). \tag{3.15}$$

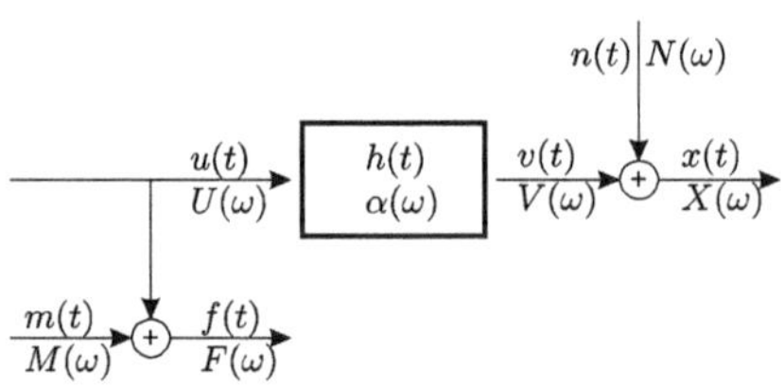

FIGURE 3.5 A block diagram for a SISO system.

For arbitrary correlations between the signals $u(t)$ and $v(t)$ and the noise $m(t)$ and $n(t)$, auto- and cross-spectral density functions (2.43) are:

$$\begin{aligned} S_{ff}(\omega) &= S_{uu}(\omega) + S_{mm}(\omega) + S_{mu}(\omega) + S_{um}(\omega), \\ S_{xx}(\omega) &= S_{vv}(\omega) + S_{nn}(\omega) + S_{nv}(\omega) + S_{vn}(\omega), \\ S_{fx}(\omega) &= S_{uv}(\omega) + S_{un}(\omega) + S_{mv}(\omega) + S_{mn}(\omega). \end{aligned} \tag{3.16}$$

Using the definition of auto- (2.77) and cross-spectra densities (2.78), we have

$$\begin{aligned} S_{vv}(\omega) &= |\alpha(\omega)|^2 \, S_{uu}(\omega), \\ S_{uv}(\omega) &= \alpha(\omega) \, S_{uu}(\omega). \end{aligned} \tag{3.17}$$

$\hat{H}_1$ estimator

Measuring the excitation $f(t)$ and response $x(t)$, the latter usually covers higher dynamical range and is therefore more sensitive to measurement noise. Due to this, let us assume that there is no input noise $m(t) = 0$ and that the output noise

$n(t)$ is uncorrelated; consequently, Eq. (3.16) simplifies to

$$\begin{aligned} S_{ff}(\omega) &= S_{uu}(\omega), \\ S_{xx}(\omega) &= S_{vv}(\omega) + S_{nn}(\omega), \\ S_{fx}(\omega) &= S_{uv}(\omega), \end{aligned} \tag{3.18}$$

and using (3.17), we have

$$S_{fx}(\omega) = \alpha(\omega)\, S_{ff}(\omega). \tag{3.19}$$

If (3.19) is used to estimate the FRF from measurements, then this is written as

$$\hat{H}_1(\omega) = \hat{\alpha}_1(\omega) = \frac{\hat{S}_{fx}(\omega)}{\hat{S}_{ff}(\omega)}, \tag{3.20}$$

where we have used the symbol ^ to denote the estimation from measurement (auto $\hat{S}_{fx}(\omega)$ and cross $\hat{S}_{ff}(\omega)$ spectra are obtained by averaging a random process).

The coherence function (2.79) becomes

$$\gamma_{fx}^2 = \frac{|S_{uv}|^2}{S_{uu}\,(S_{vv} + S_{nn})}, \tag{3.21}$$

which shows that the PSD of noise at the output S_{nn} decreases the coherence and therefore the quality of the estimated FRF.

If excitation noise $m(t)$ cannot be neglected, then $S_{ff} = S_{uu} + S_{mm}$, and

$$\hat{H}_1(\omega) = \alpha(\omega) \left(1 + \frac{\hat{S}_{mm}}{\hat{S}_{uu}}\right)^{-1}, \tag{3.22}$$

and consequently the $\hat{H}_1$ estimator underestimates the true FRF $\alpha = S_{uv}/S_{uu}$ in case of noise at the excitation side.

$\hat{H}_2$ estimator

Here we study the estimation of FRF when the excitation measurement noise cannot be neglected, but it is uncorrelated. Further, let us assume that the response measurement noise can be neglected ($n(t) = 0$) and consequently Eq. (3.16) simplifies to

$$\begin{aligned} S_{ff} &= S_{uu} + S_{mm}, \\ S_{xx} &= S_{vv}, \\ S_{fx} &= S_{uv}, \end{aligned} \tag{3.23}$$

and using (3.17), we have

$$\begin{aligned} S_{xx} &= S_{vv} \\ &= |\alpha(\omega)|^2 \, S_{uu} \\ &= \alpha(\omega)\, \alpha(\omega)^* \, S_{uu} \\ &= \alpha(\omega)\, S_{xf}, \end{aligned} \tag{3.24}$$

where $S_{xf} = \alpha(\omega)^* \, S_{uu}$, and $S_{xf} = S^*_{fx}$ was used in the last line. If (3.24) is used to estimate the FRF from measurements, then this is written as

$$\hat{H}_2(\omega) = \hat{\alpha}_2(\omega) = \frac{\hat{S}_{xx}}{\hat{S}_{xf}}. \tag{3.25}$$

The coherence function (2.79) becomes

$$\gamma^2_{fx} = \frac{|S_{fx}|^2}{(S_{uu} + S_{mm})\, S_{vv}}, \tag{3.26}$$

showing that also the noise at the excitation side decreases the coherence and therefore the quality of the estimated FRF.

Dividing the $\hat{H}_1$ estimator by $\hat{H}_2$, we obtain the coherence estimator

$$\hat{\gamma}^2_{fx} = \frac{\hat{H}_1}{\hat{H}_2}. \tag{3.27}$$

If response noise $n(t)$ cannot be neglected, then $S_{xx} = S_{vv} + S_{nn}$, and

$$\hat{H}_2(\omega) = \alpha(\omega) \left(1 + \frac{\hat{S}_{nn}}{\hat{S}_{vv}} \right), \tag{3.28}$$

and consequently the $\hat{H}_2$ estimator overestimates the true FRF α in case of noise at the response side.

3.2.2 Multiple-input multiple-output system

A dynamical system with multiple inputs $F_k(\omega)$ and multiple outputs $X_j(\omega)$ as shown in Fig. 3.6 is called a *multiple-input multiple-output system* (MIMO). The indexes k and j are used for the excitation and response locations, respectively. The frequency domain response X_j is defined as

$$X_j(\omega) = \sum_{k=1}^{P} \alpha_{jk}(\omega)\, F_k(\omega), \tag{3.29}$$

where P is the number of excitation locations. Eq. (3.29) can be written as (1.174)

$$\mathbf{X}(\omega)_{L\times 1} = \boldsymbol{\alpha}(\omega)_{L\times P}\,\mathbf{F}(\omega)_{P\times 1}, \tag{3.30}$$

where $\mathbf{X}$ and $\mathbf{F}$ are the response and excitation vectors, respectively. The response vector in frequency domain is, for example, defined as

$$\mathbf{X}(\omega) = [X_1(\omega), X_2(\omega), \ldots, X_P(\omega)]^{\mathrm{T}}, \tag{3.31}$$

and $\boldsymbol{\alpha}$ is the receptance matrix (1.175); here the dimensions are subscribed for clarity. The number of response locations is L and presents the upper limit of the number of excitation locations P.

In Fig. 3.6, measurement noise is added at the excitation side $\mathbf{M}(\omega)$ and the response side $\mathbf{N}(\omega)$; $\mathbf{U}(\omega)$ and $\mathbf{V}(\omega)$ are the true spectra of force and response, which when combined with noise $\mathbf{M}(\omega)$ and $\mathbf{N}(\omega)$ produce the measured spectra of force $\mathbf{F}(\omega)$ and response $\mathbf{X}(\omega)$ [25]:

$$\begin{aligned}\mathbf{X}(\omega) &= \mathbf{V}(\omega) + \mathbf{N}(\omega),\\ \mathbf{F}(\omega) &= \mathbf{U}(\omega) + \mathbf{M}(\omega).\end{aligned} \tag{3.32}$$

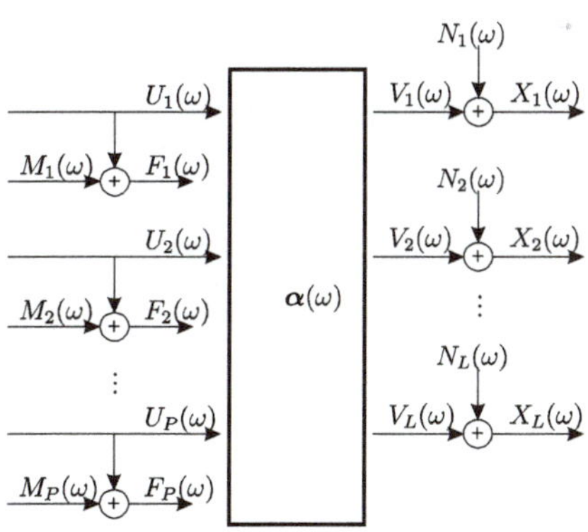

FIGURE 3.6 A block diagram of a MIMO system incorporating noise at the excitation and response side.

Similarly as in SISO system, the FRF can be estimated with estimators; however, here the receptance matrix $\boldsymbol{\alpha}(\omega)$ is estimated (1.175). Only final equations are given, and the interested reader should check Fu and He [25] for details.

$\hat{H}_1$ estimator

The $\hat{H}_1$ estimator is defined as

$$\hat{\mathbf{H}}_1(\omega)_{L\times P} = \hat{\boldsymbol{\alpha}}_1(\omega)_{L\times P} = \hat{\mathbf{S}}_{XF}(\omega)_{L\times P}\,\hat{\mathbf{S}}_{FF}(\omega)^{-1}_{P\times P}, \tag{3.33}$$

where $\hat{\mathbf{S}}_{FF}(\omega)$ needs to be inverted at each ω and is the estimator of the excitation spectral density matrix with elements at index (a,b) (2.43):

$$\hat{S}_{FF_{a,b}}(\omega) = \frac{1}{N}\sum_{i=1}^{N}\left(\frac{1}{T}\,F_{a,i}^{*}(\omega)\,F_{b,i}(\omega)\right), \qquad a,b = 1,2,\ldots P. \tag{3.34}$$

In Eq. (3.34), N repetitions of the experiment are assumed, where each experiment is denoted by index i.

Further, $\hat{\mathbf{S}}_{XF}(\omega)$ is the estimator of the excitation/response cross-spectral density matrix with elements at index (a,b):

$$\hat{S}_{XF_{a,b}}(\omega) = \frac{1}{N}\sum_{i=1}^{N}\left(\frac{1}{T}\,X_{a}^{*}(\omega)\,F_{b}(\omega)\right), \qquad a = 1,2,\ldots L, \qquad b = 1,2,\ldots P. \tag{3.35}$$

$\hat{H}_2$ estimator

The $\hat{H}_2$ estimator is defined as

$$\hat{\mathbf{H}}_2(\omega)_{L\times P} = \hat{\boldsymbol{\alpha}}_2(\omega)_{L\times P} = \hat{\mathbf{S}}_{XX}(\omega)_{L\times L}\,\hat{\mathbf{S}}_{FX}(\omega)_{L\times P}^{-1}. \tag{3.36}$$

Coherence

Ordinary coherence function for the kth excitation and the jth response is defined as

$$\gamma_{jk}^{2}(\omega) = \frac{|S_{XF_{jk}}(\omega)|^2}{S_{FF_{kk}}(\omega)\,S_{XX_{jj}}(\omega)}. \tag{3.37}$$

3.2.3 Comparison of estimators $\hat{H}_1$ and $\hat{H}_2$

Fig. 3.7 shows the comparison of estimators $\hat{H}_1$ and $\hat{H}_2$ based on an average of 25 simulated measurements with simulated Gaussian noise at the excitation (standard deviation 0.2) and the response (standard deviation 0.1) side. The estimators are plotted against the FRF obtained as the ratio of the response and the force from Eq. (1.35) for a single measurement $\hat{H}_{n=1}(\omega) = X(\omega)/F(\omega)$ and against the true FRF $H(\omega)$.

The estimators $\hat{H}_1$ and $\hat{H}_2$ produce better estimates than the single measurement $\hat{H}_{n=1}$. It can be seen that the estimator $\hat{H}_1$ underestimates the true FRF close to resonance, where the autospectral density of the true excitation $\hat{S}_{uu}$ becomes small if compared to the autospectral density of the input noise $\hat{S}_{mm}$, and consequently the ratio $\hat{S}_{mm}/\hat{S}_{uu}$ in Eq. (3.22) increases. Similarly, we can see that the estimator $\hat{H}_2$ overestimates the true FRF close to antiresonance when the autospectral density of the true response $\hat{S}_{vv}$ becomes small if compared to the autospectral density of the output noise $\hat{S}_{nn}$, and consequently the ratio $\hat{S}_{nn}/\hat{S}_{vv}$ in Eq. (3.28) increases.

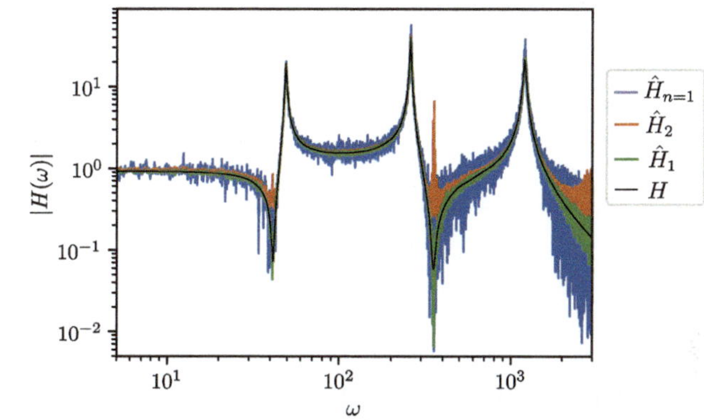

FIGURE 3.7 The comparison of the estimators $\hat{H}_1$ and $\hat{H}_2$ with a simulated H for 25 simulated measurements.

Fig. 3.8 is a *Nyquist plot* of the estimators from Fig. 3.7. A Nyquist plot is a plot of the imaginary part $\Im\{\}$ with respect to the real part $\Re\{\}$. The resonances in the Nyquist plot are seen as circles, because the phase at the resonant frequency changes from 0 to π. The Nyquist plot in Fig. 3.8 confirms that the estimator $\hat{H}_1$ underestimates, and $\hat{H}_2$ is closer to the true value.

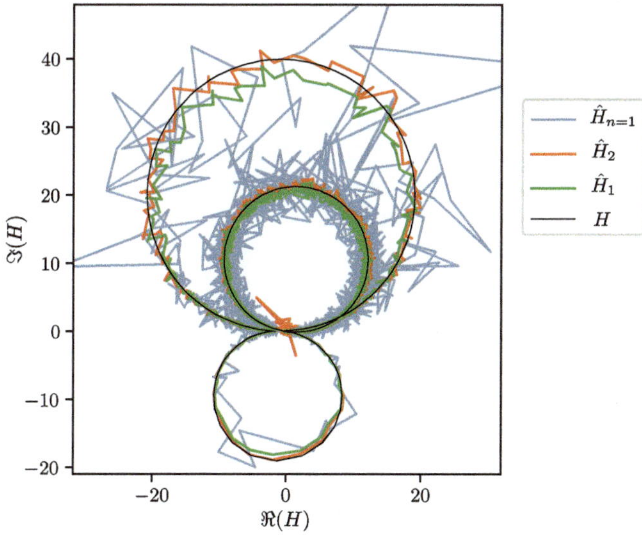

FIGURE 3.8 Nyquist plot of the simulated FRF H and the $\hat{H}_1$ and $\hat{H}_2$ estimators.

Fig. 3.9 shows the coherence for the data in Fig. 3.7.

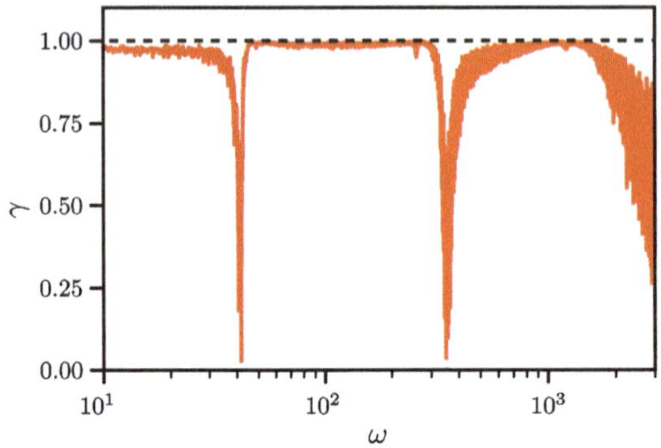

FIGURE 3.9 Coherence for the simulated data in Fig. 3.7.

3.3 Modal parameter identification

One of the main subjects of modal analysis is the identification of the modal parameters from measured data. The modal parameters in question are the eigenfrequencies, the damping, and the mode shapes, which comprise the modal model. The modal model is preferred, because it produces the most compressed and representable information of the measured system. The identified dynamics can serve as a basis for structural modifications, substructuring, model updating, fatigue predictions, damage identification, and so on.

The identification methods are based on physical models described in Chapter 1, and the modal parameters are determined as the values that minimize the reconstruction error. Different methods are based on different physical models.

With regard to the number of the degrees-of-freedom, the methods can be divided into methods based on single-degree-of-freedom (SDOF) and multi-degree-of-freedom (MDOF) models.

According to the number of FRFs and the way the measurements are performed, the methods are divided into *SISO*, *SIMO*, and *MIMO* methods. SISO stands for *Single-Input Single-Output*; this is when a single response point is measured along with a single excitation and the identification is performed on one FRF only. Often, multiple FRFs are measured either in a single measurement or by repeated measurements and by changing the location of the measured response or the location of excitation, the first is typically done with multiple accelerometers or full-field measurement methods, and the last is typically done with a *roving hammer*. If the modal parameters were to be identified on each FRF individually, the obtained identified eigenvalues would, due to measurement errors, be different for different FRFs, and this is not correct, since the eigenvalues are the same for all FRFs according to the theory. To overcome this problem, *Single-Input Multiple-Output* (SIMO) methods account multiple FRFs with a common response or excitation location in the identification and produce identified eigenvalues that represent all the measured FRFs. SIMO methods work on a single (albeit an incomplete) row or column of the FRF matrix (1.175). Due to the reciprocity (1.176) and the modal constants consistency (1.177), the displacement receptance matrix $\boldsymbol{\alpha}(\omega)$ is fully defined if one column or one row is measured; however, this is not the case for the strain receptance matrix; see Eq. (1.217) for details. Anyway, also for the displacement receptance matrix, it is preferred to measure an extended range of the FRF matrix, because when exciting at a chosen point, certain modes are excited more than others, there is also a real chance that the chosen excitation location coincides with a node point and the mode shape gets excited inadequately for a good measurement to be obtained. Additionally, an extended measured FRF matrix serves as redundant information improving the modal identification when using *Multiple-Input Multiple-Output* (MIMO) methods.

According to the domain upon which the method is based, methods are divided into time domain and frequency domain methods.

3.3.1 Complex exponential

The complex exponential method is based on an impulse response in the time domain $h(t)$ (1.79). The impulse response is a response of the structure to an unitary impulse (ideally, a Dirac impulse) and can also be obtained as the inverse of the FRF (1.91). For an MDOF viscous damping model, the FRF in partial fraction form is

$$\alpha(\omega) = \sum_{r=1}^{2N} \frac{R_r}{\mathrm{i}\,\omega - \lambda_r}. \tag{3.38}$$

The equation equals Eq. (1.189), but the form is not divided into two conjugated eigenvalues, which are now enclosed in the expanded range $2N$, so that

$$\lambda_r = \lambda_{r+N}^* \tag{3.39}$$

and

$$R_r = R_{r+N}^*. \tag{3.40}$$

The impulse response based on the model (3.38) is (see also Eq. (1.91))

$$h(t) = \sum_{r=1}^{2N} R_r\, \mathrm{e}^{\lambda_r t}. \tag{3.41}$$

The response $h(t)$ for sequential discretely sampled intervals Δt up to the last measured value at the $U+1$ time-step is according to (3.41):

$$h_0 = h(0) = \sum_{r=1}^{2N} R_r, \tag{3.42}$$

$$h_1 = h(\Delta t) = \sum_{r=1}^{2N} R_r \, e^{\lambda_r \, (\Delta t)}, \tag{3.43}$$

$$h_2 = h(2\Delta t) = \sum_{r=1}^{2N} R_r \, e^{\lambda_r \, (2\Delta t)}, \tag{3.44}$$

$$\vdots \quad \vdots \quad \vdots$$

$$h_U = h(U\,\Delta t) = \sum_{r=1}^{2N} R_r \, e^{\lambda_r \, (U\,\Delta t)}. \tag{3.45}$$

Using the symbol

$$V_r = e^{\lambda_r \, (\Delta t)}, \tag{3.46}$$

they become

$$h_0 = \sum_{r=1}^{2N} R_r, \tag{3.47}$$

$$h_1 = \sum_{r=1}^{2N} R_r \, V_r, \tag{3.48}$$

$$h_2 = \sum_{r=1}^{2N} R_r \, V_r^2, \tag{3.49}$$

$$\vdots \quad \vdots$$

$$h_U = \sum_{r=1}^{2N} R_r \, V_r^U. \tag{3.50}$$

The values V_r and R_r are unknown. First, V_r (and consequentially the eigenvalues λ_r) are determined, and secondly the modal constants R_r. The values V_r are determined using the Prony method; the eigenvalues λ_r appear in complex conjugate pairs, and so do the values V_r; therefore there exists a polynomial V_r of order U with real coefficients β_n such that

$$\beta_0 + \beta_1 \, V_r + \beta_2 \, V_r^2 + \cdots + \beta_U \, V_r^U = 0. \tag{3.51}$$

By multiplication of both sides of Eqs. (3.47)–(3.50) with β_0 up to β_U and summing these equations, we obtain the following equation:

$$\sum_{n=0}^{U} \beta_n \, h_n = \sum_{n=0}^{U} \left(\beta_n \sum_{r=1}^{2N} R_r \, V_r^n \right) = \sum_{r=1}^{2N} \left(R_r \sum_{n=0}^{U} \beta_n \, V_r^n \right). \tag{3.52}$$

The sum $\sum_{n=0}^{U} \beta_n\, V_r^n$ equals the polynomial equation (3.51), and therefore the sum equals zero, and the following expression can be written regardless of V_r:

$$\sum_{n=0}^{U} \beta_n\, h_n = 0. \tag{3.53}$$

At this point the polynomial order (3.51) is limited with U related to the number of time-steps, which can be increased up to the number $2\,N$ of eigenvalues. Further, to obtain the $2\,N$ unknown values of β_n, $2\,N$ Eqs. (3.53), each shifted for a time instance Δt, have to be written, and the last β_n value is set to unity, that is, $\beta_{2\,N} = 1$. This way we derive the following system of equations:

$$\underset{[2N\times 2N]}{\begin{bmatrix} h_0 & h_1 & \dots & h_{2N-1} \\ h_1 & h_2 & \dots & h_{2N} \\ h_2 & h_3 & \dots & h_{2N+1} \\ \vdots & \vdots & \ddots & \vdots \\ h_{2N-1} & h_{2N} & \dots & h_{4N-2} \end{bmatrix}} \underset{[2N\times 1]}{\begin{Bmatrix} \beta_0 \\ \beta_1 \\ \beta_2 \\ \vdots \\ \beta_{2N-1} \end{Bmatrix}} = - \underset{[2N\times 1]}{\begin{Bmatrix} h_{2N} \\ h_{2N+1} \\ h_{2N+2} \\ \vdots \\ h_{4N-1} \end{Bmatrix}} \tag{3.54}$$

or, in short form,

$$\mathbf{h}\,\boldsymbol{\beta} = \mathbf{h}'. \tag{3.55}$$

The coefficient β_{2N} is set to 1, because the polynomial (3.51) can be multiplied with an arbitrary constant not changing the solution.

By solving the inverse values β_n are determined for $n = 0, 1, \ldots, 2N - 1$, and $\beta_{2N} = 1$. The calculated values are then substituted into the polynomial (3.51). The roots of the polynomial indicate the values V_r. The eigenvalues λ_r are calculated from V_r via relation (3.46).

After the values V_r are determined, a system of equations based on (3.47) can be formed for time instances from h_0 up to h_{2N-1}:

$$\begin{bmatrix} 1 & 1 & \dots & 1 \\ V_1 & V_2 & \dots & V_{2N} \\ \vdots & \vdots & \ddots & \vdots \\ V_1^{2N-1} & V_2^{2N-1} & \dots & V_{2N}^{2N-1} \end{bmatrix} \begin{Bmatrix} R_1 \\ R_2 \\ \vdots \\ R_{2N} \end{Bmatrix} = \begin{Bmatrix} h_0 \\ h_1 \\ \vdots \\ h_{2N-1} \end{Bmatrix}. \tag{3.56}$$

Solving this system produces the modal constants R_r. By considering that the modal constants appear in conjugated pairs as well, merely N time instances can be used.

As the method is based on the time domain impulse response, it will not exhibit the effects of windowing when the impulse response is measured directly, which is when a modal hammer is used for excitation. If the impulse response is not measured directly, then we can obtain it from the receptance by using the inverse Fourier transform.

The CE method is popular because it does not require any initial conditions and identifies the modal parameters in two steps: first, the eigenvalues and then the modal constants. The length of the time series U determines the number of modes to be identified. To identify $2N$ eigenvalues, $U = 4N$ time instances are required.

Stabilization diagram

An appropriate order of identification is usually not known prior to the identification process. The identification process is typically repeated for variable orders of N, and the identified eigenvalues are hand-picked from the *stabilization diagram*. The stabilization diagram shows the result of successive identification iterations; see Fig. 3.10. In the figure the following color code is used to mark damping stabilization: green for stabilized damping, red for nonstabilized damping, and yellow for positive real part (unstable pole). Markers are used for eigenfrequency stabilization: cross is used for stabilized and dot for nonstabilized eigenfrequency.

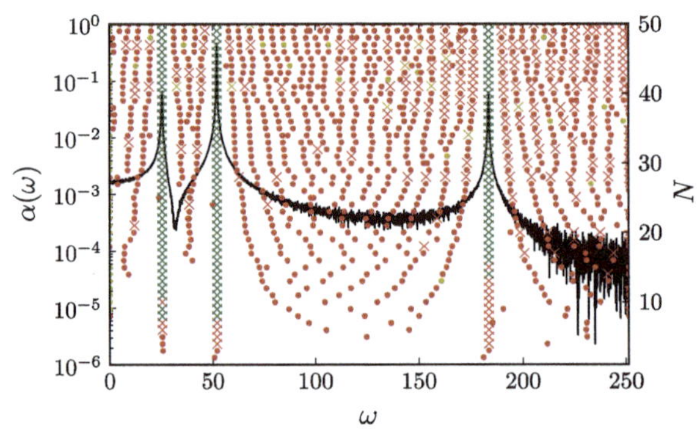

FIGURE 3.10 An example of a stabilization diagram.

The picked eigenvalues are then used in the follow-up calculation of the modal constants. This type of a procedure is very efficient and most commonly used.

3.3.2 Least squares complex exponential

The CE method is a SISO method and therefore only works on individual FRFs. The method was expanded to the SIMO case by Brown et al. [31] producing the *least squares complex exponential* (LSCE) method. Multiple outputs are integrated in the identification by adding other responses to the system of

Eqs. (3.54), thereby overdetermining the system:

$$\underset{[2\,N\,L\times 2\,N]}{\begin{bmatrix} \mathbf{h}_1 \\ \mathbf{h}_2 \\ \vdots \\ \mathbf{h}_L \end{bmatrix}} \underset{[2\,N\times 1]}{\boldsymbol{\beta}} = \underset{[2\,N\,L\times 1]}{\begin{Bmatrix} \mathbf{h}'_1 \\ \mathbf{h}'_2 \\ \vdots \\ \mathbf{h}'_L \end{Bmatrix}}. \tag{3.57}$$

The notations $\mathbf{h}_j$ and $\mathbf{h}'_j$ are set in the wake of (3.55), and the added indexes j represent the different response locations in the range $j = 1, 2, \dots, L$. The equations can be written in the short form

$$\mathbf{h}_{1\dots L}\,\boldsymbol{\beta} = \mathbf{h}'_{1\dots L}. \tag{3.58}$$

A pseudoinverse $\mathbf{h}^{\dagger}_{1\dots L}$ (see App. A) is needed to solve the overdetermined system (3.57). The solution produces results for the minimum of the sum of the squares (a least squares solution):

$$\boldsymbol{\beta} = \mathbf{h}^{\dagger}_{1\dots L}\,\mathbf{h}'_{1\dots L}. \tag{3.59}$$

The obtained $\boldsymbol{\beta}$ are then used in the same procedure as in the CE method (Section 3.3.1) to produce the identified eigenvalues of the system.

Since the system is already overdetermined, it is possible to increase the time series length from $4\,N$, thereby increasing the overdetermination and improving the identification. Additional time points are to be appended in each $[h]_j$, thereby expanding it to $2\,N^+$ $(2\,N < 2\,N^+)$ rows. The rows can be expanded up to $2\,N^+ = U - 2\,N$, where U is the length of the discrete time series:

$$\mathbf{h}_{[2\,N^+\times 2\,N]}\,\boldsymbol{\beta}_{[2\,N\times 1]} = \mathbf{h}'_{[2\,N^+\times 1]}. \tag{3.60}$$

The modal constants are determined separately for each FRF in the same way as in the CE method; this is because they are unique to the respective FRFs. As with the identification of the eigenvalues, an extended time range can be used to overdetermine system (3.56). Additional rows need to be added to system (3.56) to accommodate the extended time series:

$$\begin{bmatrix} 1 & 1 & \dots & 1 \\ V_1 & V_2 & \dots & V_{2\,N} \\ \vdots & \vdots & \ddots & \vdots \\ V_1^{2\,N^+-1} & V_2^{2\,N^+-1} & \dots & V_{2\,N}^{2\,N^+-1} \end{bmatrix} \begin{Bmatrix} {}_1R_j \\ {}_2R_j \\ \vdots \\ {}_{2\,N}R_j \end{Bmatrix} = \begin{Bmatrix} h_j(0) \\ h_j(\Delta t) \\ \vdots \\ h_j((2\,N^+ - 1)\Delta t) \end{Bmatrix}. \tag{3.61}$$

Again, a pseudoinverse has to be determined to solve the system and produce the identified modal constants

$$\mathbf{R}_j = \mathbf{V}^{\dagger}\, \mathbf{h}_j. \tag{3.62}$$

The modal constants $\mathbf{R}_j$ have to be determined for every response location j separately; however, this is not computationally demanding, because the pseudoinverse is based on the eigenvalues and is the same for all response locations, and therefore it needs to be determined only once.

A Multiple-Input Multiple-Output (MIMO) variant of the method exists and is known as the *polyreference complex exponential* (PRCE); see, for example, [4].

3.3.3 Least squares complex frequency

Least squares complex frequency (LSCF) is a frequency domain method derived from a *common denominator* receptance model [32,33]:

$$\alpha_j(\omega) = \frac{\sum_{r=0}^{2N} a_{j,r}\, \varrho_r(\omega)}{\sum_{r=0}^{2N} b_r\, \varrho_r(\omega)}. \tag{3.63}$$

In the common denominator model the values b_r governing the eigenvalues λ_r of the system are the same for each response location j (and one excitation location[2]); $\varrho_r(\omega)$ is a base function and for the case of LSCF equals $\varrho_r(\omega) = \exp(-\mathrm{i}\, r\, \Delta t\, \omega)$ [34] (derived from the discrete Z-domain).[3] Index k is omitted, because the excitation should be the same for all measured responses (SIMO measurement).

A least squares criterion is used to produce a system of equations to fit the measurement $\hat{\alpha}_j(\omega)$ to the approximation $\alpha_j(\omega)$ [32]:

$$e_{n,j} = W_j(\Omega_n) \left(\frac{\sum_{r=0}^{2N} a_{j,r}\, \varrho_r(\Omega_n)}{\sum_{r=0}^{2N} b_r\, \varrho_r(\Omega_n)} - \hat{\alpha}_j(\Omega_n) \right), \tag{3.64}$$

where $W_j(\omega)$ is the weighting function to better control the identification, index $j = 1, \ldots, L$ represents the various response points, and index $n = 1, \ldots, U$ represents the frequency points. The angular frequency Ω_n represents the elements of the frequency vector from the FRF estimator $\hat{\alpha}$, and n is the frequency vector index. By multiplying both sides with the denominator we obtain the linearized error equation

$$e'_{n,j} = W_j(\Omega_n) \sum_{r=0}^{2N} \Big(a_{j,r}\, \varrho_r(\Omega_n) - b_r\, \varrho_r(\Omega_n)\, \hat{\alpha}_j(\Omega_n) \Big). \tag{3.65}$$

---

2 For multiple input, see [33].

3 A base function $\varrho_r = (\mathrm{i}\, \omega)^r$ is used in the rational fraction polynomial (RFP) method (based on a continuous model), and for a better conditioning, the Forsythe polynomials are typically used.

The error for every frequency point is written in matrix form as

$$\mathbf{e}'_j = \left\{ \begin{array}{c} e'_{1,j} \\ e'_{2,j} \\ \vdots \\ e'_{U,j} \end{array} \right\} = \left[\begin{array}{cc} \mathbf{P}_j & \mathbf{T}_j \end{array} \right] \left\{ \begin{array}{c} \mathbf{a}_j \\ \mathbf{b} \end{array} \right\}, \tag{3.66}$$

where

$$\mathbf{P}_j = \left[\begin{array}{cccc} W_j(\Omega_1)\,\varrho_0(\Omega_1) & W_j(\Omega_1)\,\varrho_1(\Omega_1) & \dots & W_j(\Omega_1)\,\varrho_{2N}(\Omega_1) \\ W_j(\Omega_2)\,\varrho_0(\Omega_2) & W_j(\Omega_2)\,\varrho_1(\Omega_2) & \dots & W_j(\Omega_2)\,\varrho_{2N}(\Omega_2) \\ \vdots & \vdots & \ddots & \vdots \\ W_j(\Omega_U)\,\varrho_0(\Omega_U) & W_j(\Omega_U)\,\varrho_1(\Omega_U) & \dots & W_j(\Omega_U)\,\varrho_{2N}(\Omega_U) \end{array} \right], \tag{3.67}$$

$$\mathbf{T}_j = - \left[\begin{array}{ccc} W_j(\Omega_1)\,\tilde{\alpha}_j(\Omega_1)\,\varrho_0(\Omega_1) & \dots & W_j(\Omega_1)\,\tilde{\alpha}_j(\Omega_1)\,\varrho_{2N}(\Omega_1) \\ W_j(\Omega_2)\,\tilde{\alpha}_j(\Omega_2)\,\varrho_0(\Omega_2) & \dots & W_j(\Omega_2)\,\tilde{\alpha}_j(\Omega_2)\,\varrho_{2N}(\Omega_2) \\ \vdots & \ddots & \vdots \\ W_j(\Omega_U)\,\tilde{\alpha}_j(\Omega_U)\,\varrho_0(\Omega_2) & \dots & W_j(\Omega_U)\,\tilde{\alpha}_j(\Omega_U)\,\varrho_U(\Omega_2) \end{array} \right], \tag{3.68}$$

and $\mathbf{a}_j$ and $\mathbf{b}$ are the vectors of the numerator and denominator polynomial coefficients:

$$\mathbf{a}_j = \left\{ \begin{array}{c} a_{j,0} \\ a_{j,1} \\ \vdots \\ a_{j,2N} \end{array} \right\}, \tag{3.69}$$

$$\mathbf{b} = \left\{ \begin{array}{c} b_0 \\ b_1 \\ \vdots \\ b_{2N} \end{array} \right\}. \tag{3.70}$$

To combine all response points $j = 1, \dots, L$ along with all frequency points $n = 1, \dots, U$ in the identification, the error is produced as

$$
\begin{aligned}
\ell &= \sum_{j=1}^{L}\sum_{n=1}^{U} |e'_{n,j}|^2 \\
&= \sum_{j=1}^{L} \left(\mathbf{e}'^{\mathrm{H}}_j \mathbf{e}'_j\right) \\
&= \sum_{j=1}^{L} \left\{ \begin{array}{c} \mathbf{a}_j \\ \mathbf{b} \end{array} \right\}^{\mathrm{T}} \left[\begin{array}{cc} \mathbf{R}_j & \mathbf{S}_j \\ \mathbf{S}_j^{\mathrm{T}} & \mathbf{O}_j \end{array} \right] \left\{ \begin{array}{c} \mathbf{a}_j \\ \mathbf{b} \end{array} \right\},
\end{aligned} \tag{3.71}
$$

where

$$\mathbf{R}_j = \Re\left\{\mathbf{P}_j^{\mathrm{H}} \mathbf{P}_j\right\}, \tag{3.72}$$

$$\mathbf{S}_j = \Re\left\{\mathbf{P}_j^{\mathrm{H}} \mathbf{T}_j\right\}, \tag{3.73}$$

$$\mathbf{O}_j = \Re\left\{\mathbf{T}_j^{\mathrm{H}} \mathbf{T}_j\right\}. \tag{3.74}$$

The elements of the matrices $\mathbf{O}_j$, $\mathbf{S}_j$, $\mathbf{R}_j$ (l is row, c is column, $n = 1, 2, \ldots, U$ are frequency points) are determined by

$$R_j(l,c) = \Re\left\{ \sum_{n=1}^{U} |W_j(\Omega_n)|^2 \, \varrho^*_{l-1}(\Omega_n)\, \varrho_{c-1}(\Omega_n) \right\}, \tag{3.75}$$

$$S_j(l,c) = -\Re\left\{ \sum_{n=1}^{U} |W_j(\Omega_n)|^2 \, H_j(\Omega_n)\, \varrho^*_{l-1}(\Omega_n)\, \varrho_{c-1}(\Omega_n) \right\}, \tag{3.76}$$

$$O_j(l,c) = \Re\left\{ \sum_{n=1}^{U} |W_j(\Omega_n)\, H_j(\Omega_n)|^2 \, \varrho^*_{l-1}(\Omega_n)\, \varrho_{c-1}(\Omega_n) \right\}. \tag{3.77}$$

Since $\varrho_r(\omega) = \mathrm{e}^{-\mathrm{i}\, r\, \Delta t\, \omega}$ and $\omega_p = 2\pi\, p/(\Delta t\, U)$, the elements take the form

$$R_j(l,c) = \Re\left\{ \sum_{n=1}^{U} |W_j(\Omega_n)|^2 \, \mathrm{e}^{2\pi\, \mathrm{i}\, (l-c)\, n/U} \right\}, \tag{3.78}$$

$$S_j(l,c) = -\Re\left\{ \sum_{n=1}^{U} |W_j(\Omega_n)|^2 \, \tilde{\alpha}_j(\Omega_n)\, \mathrm{e}^{2\pi\, \mathrm{i}\, (l-c)\, n/U} \right\}, \tag{3.79}$$

$$O_j(l,c) = \Re\left\{ \sum_{n=1}^{U} |W_j(\Omega_n)\, \tilde{\alpha}_j(\Omega_n)|^2 \, \mathrm{e}^{2\pi\, \mathrm{i}\, (l-c)\, n/U} \right\}. \tag{3.80}$$

We can see that Eqs. (3.78)–(3.80) equal the discrete Fourier transforms (2.5) of the functions $|W_j(\Omega_n)|^2$, $|W_j(\Omega_n)|^2\, \tilde{\alpha}_j(\Omega_n)$, and $|W_j(\Omega_n)\, \tilde{\alpha}_j(\Omega_n)|^2$, and therefore the fast Fourier algorithm can be used for a fast calculation [32]. Further, the matrices are of a Toeplitz structure, and the elements l, c depend on $l - c$ only. With repetitive elements, the matrix $\mathbf{S}_j$ is, for example, defined as

$$\mathbf{S}_j = \begin{bmatrix} S_j(0) & S_j(-1) & \dots & S_j(-2N) \\ S_j(1) & S_j(0) & \dots & S_j(-2N+1) \\ \vdots & \vdots & \ddots & \vdots \\ S_j(2N) & S_j(2N-1) & \dots & S_j(0) \end{bmatrix}, \tag{3.81}$$

where

$$S_j(r) = -\Re\left\{ \sum_{n=1}^{U} |W_j(\Omega_n)|^2 \, \tilde{\alpha}_j(\Omega_n) \, \mathrm{e}^{2\pi \, \mathrm{i} \, r \, n/U} \right\}. \tag{3.82}$$

The numerator and denominator coefficients $\mathbf{a}_j$, $\mathbf{b}$ are determined as the parameters that produce the minimal error of reconstruction ℓ. The minimum location is determined from the derivatives of the error function according to the coefficients $\mathbf{a}_j$ and $\mathbf{b}$:

$$\frac{\partial \ell}{\partial a_j} = 2\left(\mathbf{R}_j \, \mathbf{a}_j + \mathbf{S}_j \, \mathbf{b}\right) = 0, \tag{3.83}$$

$$\frac{\partial \ell}{\partial b} = 2\left(\sum_{j=1}^{L} \left(\mathbf{S}_j^{\mathrm{T}} \, \mathbf{a}_j + \mathbf{O}_j \, \mathbf{b}\right)\right) = 0. \tag{3.84}$$

From (3.83) $\mathbf{a}_j$ are expressed as

$$\mathbf{a}_j = -\mathbf{R}_j^{-1} \, \mathbf{S}_j \, \mathbf{b}, \tag{3.85}$$

and substituting them into (3.84), we obtain

$$\left(\sum_{j=1}^{L} \mathbf{O}_j - \mathbf{S}_j^{\mathrm{T}} \, \mathbf{R}_j^{-1} \, \mathbf{S}_j\right) \mathbf{b} = \mathbf{\Lambda} \, \mathbf{b} = 0. \tag{3.86}$$

By imposing $b_{2N} = 1$ the parameter redundancy is removed, and the last column of $\mathbf{\Lambda}$ can be shifted to the right side of the equation allowing for an inverse to be calculated producing the coefficients $\mathbf{b}$. After the coefficients $\mathbf{b}$ are known, they can be used to determine the numerator coefficients $\mathbf{a}_j$ from (3.85) for every response location j separately.

The numerator $\mathbf{a}_j$ coefficients carry the information of mode magnitude (modal constants), and the denominator $\mathbf{b}$ coefficients carry the eigenvalues information. Identified values $\mathbf{a}_j$ and $\mathbf{b}$ are produced separately in succession, and therefore the stabilization procedure from Section 3.3.1 is typically used to pick the true eigenvalues.

A Multiple-Input Multiple-Output (MIMO) variant of the method also exists. The MIMO variant is known as *poly-reference least squares complex frequency domain* (poly-LSCF or sometimes PolyMAX), for details, see [26].

3.3.4 Least squares frequency domain

Although the LSCF method can produce both the eigenvalues and the modal constants, only the eigenvalue identification is typically used, followed by the identification of modal constants with an alternate method, known as *least squares frequency domain* (LSFD) [26,35]. LSFD is an identification method that works by solving a nonlinear system of equations in an optimization procedure starting with eigenvalue estimates; however, in the case that the eigenvalues are already identified, LSFD reduces to a linear problem in the unknown modal constants. The identification is based on the frequency model from Eq. (1.189). To improve on the identification, the residuals R_L and R_U from Section 1.2.6 are incorporated [36]:

$$\alpha_{jk}(\omega) = \sum_{r=1}^{N} \left(\frac{{}_rR_{jk}}{\mathrm{i}\,\omega - \lambda_r} + \frac{{}_rR_{jk}^*}{\mathrm{i}\,\omega - \lambda_r^*} \right) - \frac{R_L}{\omega^2} + R_U. \tag{3.87}$$

By using the identified eigenvalues $\hat{\lambda}_r$ (obtained from LSCF or any other method), L equations can be written for all frequency points $n = 1, 2 \dots L$ using the measured values $\hat{\alpha}_{jk}(\Omega_n)$ in place of $\alpha_{jk}(\Omega_n)$:

$$\begin{Bmatrix} \hat{\alpha}_{jk}(\Omega_1) \\ \hat{\alpha}_{jk}(\Omega_2) \\ \vdots \\ \hat{\alpha}_{jk}(\Omega_L) \end{Bmatrix} = \begin{bmatrix} \frac{1}{\mathrm{i}\,\Omega_1 - \hat{\lambda}_1} & \frac{1}{\mathrm{i}\,\Omega_1 - \hat{\lambda}_2} & \cdots & \frac{1}{\mathrm{i}\,\Omega_1 - \hat{\lambda}_{2N}} & -\frac{1}{\Omega_1^2} & 1 \\ \frac{1}{\mathrm{i}\,\Omega_2 - \hat{\lambda}_1} & \frac{1}{\mathrm{i}\,\Omega_2 - \hat{\lambda}_2} & \cdots & \frac{1}{\mathrm{i}\,\Omega_2 - \hat{\lambda}_{2N}} & -\frac{1}{\Omega_2^2} & 1 \\ \vdots & \vdots & \ddots & \vdots & \vdots & \\ \frac{1}{\mathrm{i}\,\Omega_L - \hat{\lambda}_1} & \frac{1}{\mathrm{i}\,\Omega_L - \hat{\lambda}_2} & \cdots & \frac{1}{\mathrm{i}\,\Omega_L - \hat{\lambda}_{2N}} & -\frac{1}{\Omega_L^2} & 1 \end{bmatrix} \begin{Bmatrix} {}_1R_{jk} \\ {}_2R_{jk} \\ \vdots \\ {}_{2N}R_{jk} \\ R_L \\ R_U \end{Bmatrix}. \tag{3.88}$$

Note that the eigenvalues above N are the conjugate pairs (1.183). The system is overdetermined, and therefore a pseudoinverse (App. A) has to be used to produce the modal constants ${}_rR_{jk}$ and residues R_L in R_U.

Using LSCF to identify the eigenvalues followed by LSFD to identify the modal constants from the eigenvalues picked from stabilization diagrams is the preferred identification procedure in experimental modal analysis [26].

Chapter 4

Uniaxial vibration fatigue

Dynamically loaded structures are exposed to material fatigue, which is usually described by formation of slip bands within a grain of material when the local shear stress exceeds a threshold. With repeated loading, these bands can grow, and a crack initiates [7,37]. The goal of fatigue analysis is to estimate the time-to-failure on different points on the structure and by that also obtain the location of the critical point on the structure where the crack is estimated to occur. Such undertaking is complicated by the many factors influencing the durability and the random phenomena governing the physical process of fatigue. Estimates resulting from analysis are bound to show large scatter and must be evaluated with great care.

The undesired consequences of vibration fatigue are relevant for structures excited by random loads, which often originate from environmental processes. These can be buildings, wind turbines, and other structures, excited by wind [38], offshore installations [39], railway and road vehicles [40,41], aerospace structures [42–44], and even tools used in the manufacturing processes [45].

The stress-life approach is used for the purpose of vibration fatigue analysis. It is one of the oldest approaches, dating back to the 19th century when Wöhler [46,37] studied fatigue of rail road car axes. The damage accumulation is contributed to the stress cycles in material, and the severity of each cycle is judged against the S-N curve. The stress-life model assumes that deformations are at all times within the elastic region and that the number of cycles to failure is large (i.e., generally more than 10^4), which is why this type of material deterioration is also referred to as the high-cycle fatigue. The outlined assumptions led to the vibration fatigue analysis where the frequency range can reach as high as 2000 Hz (and beyond) and a large number of cycles accumulates in a relatively short amount of time. Another argument for designing for long time-to-failure is that it coincides with the concept of the safe-life philosophy [7,47], commonly adopted in certain industry branches (e.g., automotive), for the components designed so that they will be retired once their planned life is reached, regardless of the condition.

The highlighted stress-life approach is concerned primarily with the crack initiation part of the fatigue process, which can account for up to 80% of the time-to-failure [7]. It does not, in contrast with the fracture mechanics theory, consider the crack growth rate or planned service intervals. Rather, the number of cycles is linked to the initiation of a crack, its size on the order of 0.01 to 0.1

Vibration Fatigue by Spectral Methods. https://doi.org/10.1016/B978-0-12-822190-7.00011-6

mm [48], disregarding the more fine grained properties of the stress state and material in the vicinity of the crack tip. As such, the vibration fatigue analysis basing on the stress-life approach is useful for locating the critical points on the structure, obtaining the approximate absolute time-to-failure and, by that, rapid comparison and improvement of structural design variants.

4.1 Stress model

The stress model assumes that the loads in material do not cause plastic deformation, but on the local microscale, within the (un)favorably oriented material grains, the elastic deformation can be exceeded, and thus damage is introduced into material [7,48]. The central tool for the application of the stress fatigue model is the S-N curve, which describes the resistance of material to fatigue and gives the number N of cycles to failure for cyclic stress load of constant amplitude s. Higher stresses mean the lower number of cycles and vice versa. The law is defined by the S-N curve, also called the Wöhler curve:

$$s^k\, N = C. \tag{4.1}$$

The exponent k and coefficient C are determined through the time-consuming experimental tests on smooth tubular specimens for a specific type of metal. It is customary to present the S-N curve as a straight line in the logarithmic scale as in Fig. 4.1.

Frequently, the S-N curve is also defined with the Basquin equation [7]

$$s = s_1\, N^b, \tag{4.2}$$

where s_1 is the applied stress amplitude at one-cycle life, and b is the slope of the log-log S-N curve. The relations between different approaches to the S-N curve are

$$b = -\frac{1}{k}, \qquad s_1 = C^{\frac{1}{k}}. \tag{4.3}$$

The stress load cycle is defined by its mean s_m and amplitude s_a, as shown in Fig. 4.2, or sometimes by the range $s_r = 2\, s_a$, that is, for a uniaxial or equivalent uniaxial stress, in the form of a scalar function of time. For the case of a multiaxial stress state, some special methods, to be described in a later chapter, must first be employed, yielding the equivalent uniaxial stress.

The mean stresses can have significant impact on the load severity. The opening of the microcracks in tension effectively shortens the time-to-failure, whereas in compression they are being closed, which positively affect the time-to-failure. A number of different approaches can be used to model the mean stress effects (Gerber [49], Haigh [50], Soderberg [51], or Goodman [52]). If S_u is the ultimate tensile strength of material, then the effective cycle amplitude can

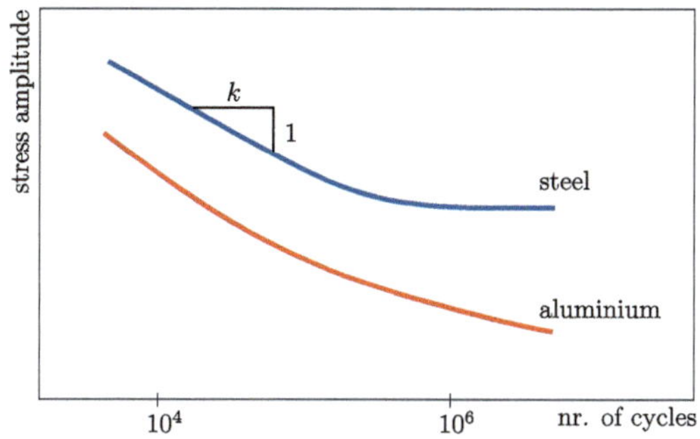

FIGURE 4.1 S-N curves for steel and aluminium.

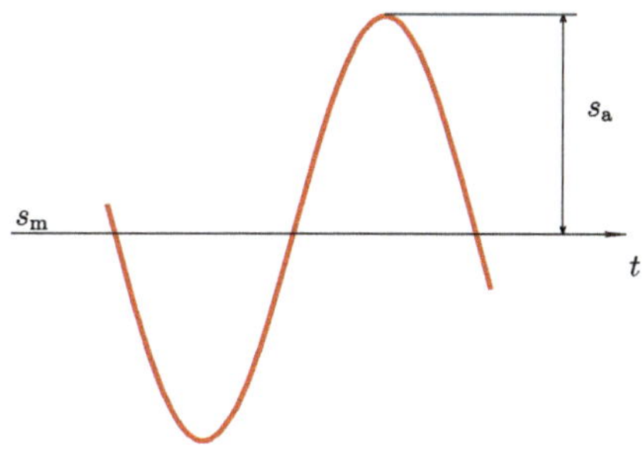

FIGURE 4.2 Definition of stress load cycle mean s_{m} and amplitude s_{a} values.

be defined as [52]

$$s_{\mathrm{eq}} = \frac{s_{\mathrm{a}}}{1 - s_{\mathrm{m}}/S_{\mathrm{u}}}. \tag{4.4}$$

The steel S-N curve tends to get flat for very high number of cycles, which is what spawned the term fatigue limit. Certain type of methods are based on the assumption that the time-to-failure is infinite for all practical purposes when the stresses drop below the endurance limit s_∞:

$$N = \begin{cases} C\, s^{-k}, & s > s_\infty, \\ \infty, & s \leq s_\infty. \end{cases} \tag{4.5}$$

These (infinite-life) approaches separate the structure into areas that are above or below this threshold. By definition such methods cannot provide the time-to-failure, at least, not directly. Further, experimental tests on machines capable of very large number of cycles show that in reality there is no fatigue limit [53]. Due to these facts, we will use henceforth the definition of the S-N curve from Eq. (4.1), not assuming fatigue limit.

4.2 Variable amplitude loads

The cycles are repeated at a constant amplitude within the bounds of a well-defined fatigue test. In a real-case scenario, the load generally consists of cycles with different amplitudes. Thus the damage contributions caused by those cycles must be combined in a systematic way.

An algorithm such as level-crossing, range-pairs, or rainflow counting is used to identify cycle amplitude distribution; see Section 2.2.7. Based on the hypothesis of damage accumulation, the contributions due to damaging loading cycles are then summed up to produce the final estimate of damage caused by the analyzed equivalent uniaxial load.

4.2.1 Damage accumulation

The hypothesis of linear damage accumulation, devised by Palmgren [54] and Miner [55], is an established tool for adding up the contributions of the damaging cycles for the stress-life models. Each cycle is ascribed to a damage portion D_i, proportional to the number of cycles-to-failure $N(s_a)$ (assumed at constant cyclic loading), given the stress amplitude of the respective damaging cycle s_a:

$$D = \sum_{i=1}^{N} D_i = \sum_{i=1}^{N} \frac{n_i(s_a)}{N(s_a)}, \tag{4.6}$$

where D is the total damage, $N(s_a)$ is determined from the S-N curve, and n_i is the total number of cycles at amplitude s_a.

The total damage at failure is often defined as $D = 1$. It depends largely on loading type, material properties, and other circumstances. There is compelling research by Eulitz and Kotte [56] showing that D can take a value within a very large interval spanning multiple orders of magnitude from 0.01 to 10, where $D < 1$ for 90% of cases.

The Palmgren–Miner hypothesis, underpinning the stress-life approach, is based on some assumptions that should be considered thoughtfully in the course of the vibration fatigue analysis. Load sequence and interaction events that have a measurable effect on time-to-failure [57] are overlooked by using the linear hypothesis. The effect of the rate of damage accumulation, dependent on the loading frequency components, is also neglected. Nevertheless, the hypothesis of linear damage accumulation remains a key part of the fatigue analysis procedure.

4.2.2 Cycle-counting methods

Counting algorithms, such as range-pairs or rainflow [58], address the task of the identification of cycle amplitudes within the time history of a random process. They operate on reversals, that is, peaks and valleys within the time history, truncating the intermediate points. The amplitudes are collected according to a set of specified rules. The cycle amplitude distribution obtained this way and the required computational effort vary depending on the selected algorithms.

Here we will only consider the rainflow counting algorithm, first proposed by Matsuishi and Endo [59]. This is a widely recognized cycle-counting algorithm, which identifies the cycles corresponding to the stress-strain hystereses as shown in Fig. 4.3. This fact is one of the main reasons for its widespread use despite the somewhat complex definition of the method.

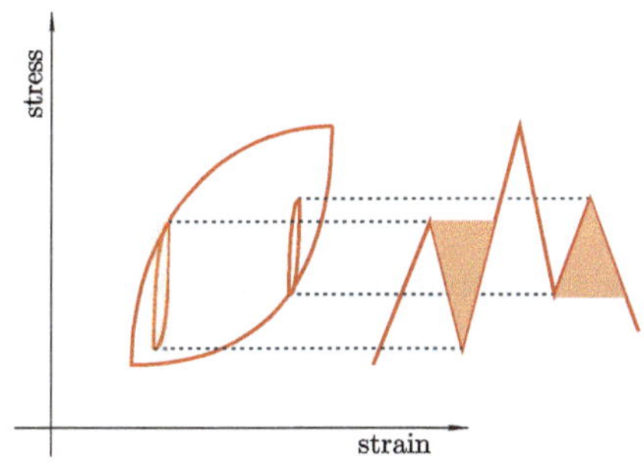

FIGURE 4.3 Connections between the rainflow-cycle amplitudes and stress-strain hystereses in material [60].

Even with rainflow, there are multiple options of how to approach the cycle counting.

Matsuishi and Endo [59] originally proposed the pagoda roof method. Algorithms that are easier to understand and implement were later suggested, namely the *3-point algorithm* by Downing and Socie [61], also listed in the ASTM standard [58], and the *4-point algorithm* by Amzallag et al. [62]. Another approach worth mentioning is the (first) nonrecursive definition given by Rychlik [63,64].

We briefly describe the 4-point algorithm. It is an iterative approach, which should be applied on the complete time history of the loading. The algorithm consists of four steps, and the results of the counting are illustrated in Fig. 4.4.

1. The time history is reduced to reversals (peaks and valleys), and intermediate points are discarded.
2. First four reversal amplitudes in the signal are taken as S_1, S_2, S_3, and S_4.
3. Inner and outer stress ranges are calculated as $\Delta S_{\mathrm{I}} = |S_2 - S_3|$ and $\Delta S_{\mathrm{O}} = |S_1 - S_4|$.
   - If the stress values of inner range points S_2 and S_3 are within the range defined by outer range points S_1 and S_4 and $\Delta S_{\mathrm{I}} \leq \Delta S_{\mathrm{O}}$, then a full cycle is counted, and points S_2 and S_3 are removed. Counting then proceeds from step 2, with the same starting point as in the previous count, that is, points S_1, S_4, S_5, and S_6 are taken into observation.
   - If the condition stated is false, then the observation shifts for one point forward in time to points S_2 to S_5, and the counting continues from step 3.

4. The points that are left at the end constitute the residue. One possible way of evaluating the residue is to count half-cycles that are also considered for damage accumulation.

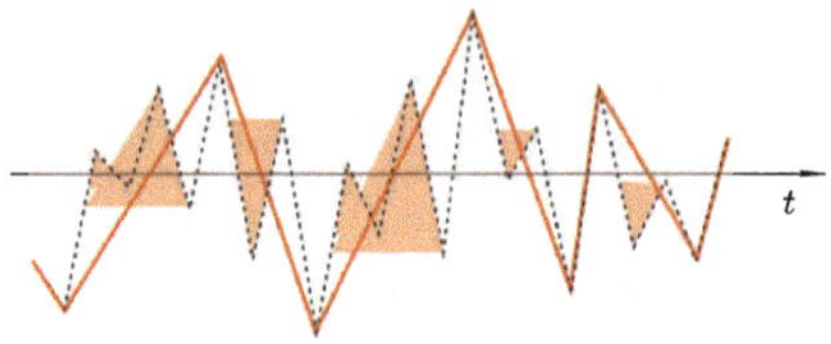

FIGURE 4.4 Identified cycle amplitudes and the residual half-cycles resulting from a rainflow count [65].

The results of the rainflow counting are commonly visualized in the form of 2D or 3D (with averages) histogram. For vibration fatigue and random process analysis, it therefore makes sense to give the cycle amplitudes in the form of a probability density function.

4.2.3 Probabilistic approach

The classical definition of the Palmgren–Miner hypothesis, written further again for clarity, deals with deterministic cycle amplitudes. Here we present it in a form that is better suited for loadings, where a variable amplitude is expected, that is, each cycle is considered separately instead of taking a group of cycles with equal amplitude:

$$D(T) = \sum_{i=1}^{N(T)} \frac{n_i}{N(s_i)} = \sum_{i=1}^{N(T)} \frac{s_i^k}{C}. \tag{4.7}$$

There exists an established formulation of the Palmgren–Miner hypothesis, which is better suited to random fatigue analysis, where expected quantities like expected peak frequency ν_{p} (2.66) and peak amplitudes PDF $p_{\mathrm{p}}(s)$ (2.67) are commonly used, rather than specific scalar values. Then we describe the damage accumulation procedure using integration, the result being the intensity of the damage instead of the absolute value D. The intensity d can be described as the damage rate or damage per unit time:

$$d = \frac{D}{T}, \tag{4.8}$$

where T is the time after which the failure occurs, and D is the absolute amount of damage that denotes failure. Oftentimes, $D = 1$ is chosen although this is not always the best choice as the actual value can widely vary.

Taking the cycle amplitude PDF $p_a(s)$ and Eqs. (4.7)–(4.8), the damage intensity can be derived by means of the integral over the stress-cycle amplitudes:

$$d = \nu_p \, C^{-1} \int_0^{\infty} s^k \, p_a(s) \, ds, \tag{4.9}$$

where ν_p is the expected peak frequency. The cycle amplitude PDF p_a is the unknown and a key factor considered by frequency-domain methods for fatigue analysis. Once p_a is defined, the time-to-failure can be determined in a straightforward way with the help of Eq. (4.8). The estimate of the peak amplitude distribution can be obtained in the frequency domain using numerical models known as the spectral moments methods [66,67].

4.3 Spectral moments methods[1]

Spectral moments methods are numeric models that can estimate the cycle amplitude probabilities not from the time history but from the frequency-domain definition of a random process. This makes it possible to perform a very comprehensive and numerically efficient analysis of fatigue on such processes [68–73]. A more classical evaluation of a random process in the time domain would require simulating an ensemble of time histories, which can be very computationally demanding.

The cycle amplitude probability estimate, usually implicitly determined using spectral moments methods, is based on the assumptions of Gaussianity of the process and its PSD. Using this kind of approach to analysis encompasses all possible realizations of the random process time history. The estimate is very fast to compute, orders of magnitude faster in comparison to the time-domain approach [66,74].

The spectral moment methods are mostly empirical models based on simulations due to the specifics of the definition of the rainflow counting algorithm and the consequent intricacies of the relationships between the PSD and rainflow cycles. Although the peak amplitude distribution and frequency can be derived in a relatively straightforward way for the normally distributed random process, the rainflow cycles cannot.

The spectral moment methods given in the following sections build up from the simple narrowband approximation to the generally applicable methods that can be used on an arbitrary broadband, normally distributed random process, to obtain a reliable time-to-failure estimate in the frequency domain.

Zero-mean process is assumed at all times, but there are available methods that can be used to compensate for that by using the established mean-stress correction models [75]. However, accounting for the respective cycle mean values is not yet possible in a straightforward way.

1 All spectral moments methods discussed in this section are implemented in open-source Python package FLife; see https://github.com/ladisk/FLife.

4.3.1 Narrowband approximation

There is a straightforward way to estimate the cycle amplitude probability density for a random process that is decidedly narrowband. This is due to the fact that all the peaks appear above the abscissa and all the valleys appear below it. This way the cycle amplitude distribution is Rayleigh, and the damage intensity, as given by Miles [76,109] is

$$d^{\mathrm{NB}} = \nu_0^+ C^{-1} \left(\sqrt{2 m_0}\right)^k \Gamma\left(1 + \frac{k}{2}\right), \tag{4.10}$$

where ν_0^+ is the expected frequency of positive zero-crossings (2.62) and is (for a narrowband process) very close to the expected peak frequency ν_{p}, C and k are material fatigue parameters defining the S-N curve (4.1), m_0 is the spectral moment (2.47), and Γ is the Euler gamma function:

$$\Gamma(z) = \int_0^\infty t^{z-1}\, \mathrm{e}^{-t}\, \mathrm{d}t. \tag{4.11}$$

This approach is only feasible when the loading process is narrowband. However, it can also be a useful tool for fatigue analysis as it always overestimates the damage intensity for a given broadband random process [77].

4.3.2 Wirsching–Light

The spectral method by Wirsching and Light [22] adapts the narrowband approximation using an empirical correction factor:

$$d^{\mathrm{WL}} = \rho_{\mathrm{WL}}\, d^{\mathrm{NB}}, \tag{4.12}$$

where d^{NB} is the narrowband approximation (4.10). The correction factor ρ_{WL} is defined as

$$\rho_{\mathrm{WL}} = a(k) + [1 - a(k)]\, (1 - \epsilon)^{b(k)}, \tag{4.13}$$

where $a(k)$ and $b(k)$ were determined as a best fit in numerical simulations for $k = 3, 4, 5$, and 6, yielding [22]

$$a(k) = 0.926 - 0.033\, k, \qquad b(k) = 1.587\, k - 2.323. \tag{4.14}$$

The spectral width parameter ϵ (4.13) is defined as [22] (see also Eq. (2.50))

$$\epsilon = \sqrt{1 - \alpha_2^2}. \tag{4.15}$$

4.3.3 $\alpha_{0.75}$ method

A simple yet effective approach was proposed by Benasciutti [78] in 2006, which uses the coefficient $\alpha_{0.75}$ to correct the narrowband approximation from Eq. (4.10):

$$d^{\mathrm{AL}} = \alpha_{0.75}^2 \, d^{\mathrm{NB}}. \tag{4.16}$$

The coefficient $\alpha_{0.75}$ (2.49) approximately compensates the conservativeness of the narrowband approximation when used on broadband processes.

4.3.4 Gao–Moan

Gao and Moan [79] presented a special method in 2008, which estimates the cycle probabilities by combining three narrowband process. It can also be generalized for an arbitrary broad band process, which has to be split into three parts beforehand.

The damage intensity d^{GM} is th sum of contributions by three narrowband random processes: high-frequency $R_H(t)$, intermediate frequency $R_M(t)$, and low-frequency process $R_L(t)$. Each of them is Rayleigh-distributed with the respective variance of $\sigma_H^2, \sigma_M^2, \sigma_L^2$. The damage intensity is

$$d^{\mathrm{GM}} = d_P + d_Q + d_H, \tag{4.17}$$

where the damage intensity d_H is due to the R_H process and is calculated using the narrowband approximation (4.10), and d_P and d_Q are intensities due to processes $R_P(t)$ and $R_Q(t)$:

$$\begin{aligned} R_P(t) &= R_H(t) + R_M(t), \\ R_Q(t) &= R_H(t) + R_M(t) + R_L(t). \end{aligned} \tag{4.18}$$

The damage accumulation model (4.9) is used to calculate d_P and d_Q, where the expected peak frequency is substituted by $\nu_{0,P}^+$ and $\nu_{0,Q}^+$ according to

$$\nu_{0,P}^+ = \sqrt{m_{2,H}\,\epsilon_H^2 + m_{2,M}\,\frac{\sigma_M}{\left(\sigma_H^2 + \sigma_M^2\right)}}, \tag{4.19}$$

$$
\begin{aligned}
\nu_{0,Q}^{+} = & \sqrt{m_{2,H}\,\epsilon_H^2 + m_{2,M}\,\epsilon_M^2 + m_{2,L}} \\
& \times \left[\frac{2\sigma_L \sqrt{\sigma_H^2 + \sigma_M^2 + \sigma_L^2} - \pi\,\sigma_H\,\sigma_M}{2\left(\sqrt{\sigma_H^2 + \sigma_M^2 + \sigma_L^2}\right)^3} \right. \\
& \left. + \frac{2\sigma_H\,\sigma_M \arctan\left(\frac{\sigma_H\,\sigma_M}{\sigma_L\sqrt{\sigma_H^2+\sigma_M^2+\sigma_L^2}}\right)}{2\left(\sqrt{\sigma_H^2 + \sigma_M^2 + \sigma_L^2}\right)^3} \right].
\end{aligned}
\tag{4.20}
$$

Here ϵ_H and ϵ_M are the Vanmarcke parameters (2.51) for processes $R_H(t)$ and $R_M(t)$. The cycle amplitude probabilities are derived for a combination of two and three narrowband random processes, for $R_P(t)$ and $R_Q(t)$, respectively, by means of the convolution integrals:

$$
p_{a,HM}(s) = \int_0^{\infty} p_{a,H}(S)\,p_{a,M}(s-S)\,\mathrm{d}S, \tag{4.21}
$$

$$
p_{a,HML}(s) = \int_0^{\infty} p_{a,HM}(S)\,p_{a,L}(s-S)\,\mathrm{d}S, \tag{4.22}
$$

where $p_{a,H}$, $p_{a,M}$, and $p_{a,L}$ are the cycle amplitude PDFs for the respective narrowband random processes. Gao and Moan used Hermite integration to perform the convolution and obtain the PDF for the combination of Rayleigh random variables [79].

It is possible to generalize the trimodal approach to an arbitrary broadband random process, which must first have its spectrum split into three parts according to some criterion. The authors [79] suggest the division using criterion of equal area under the PSD curve. Then each part takes the role of one of the processes $R_H(t)$, $R_M(t)$, and $R_L(t)$. The spectral method can be developed further by using more than three narrowband processes, at the cost of more time-expensive computation of the convolution integrals.

4.3.5 Dirlik method

The Dirlik method [73] is probably the most used spectral method. It was devised in 1985 and models the cycle amplitude probability by combining one exponential and two Rayleigh probability densities. The parameters of these PDFs are set based on numerical simulations of processes from two different PSDs shown in Fig. 4.5. All the spectra used in simulations were prepared in such a way that the ν_0^+ (2.62) and σ_x^2 (2.37) were always equal. The α_2 parameter (2.49) was changed in equidistant intervals ranging from 0 to 1.

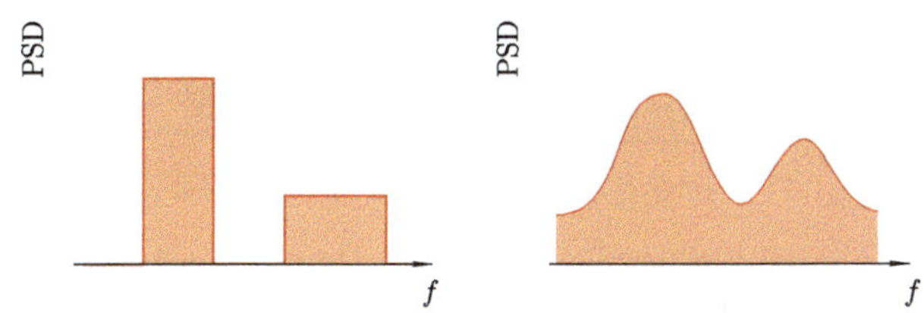

FIGURE 4.5 Shapes of spectra used in simulations to determine the coefficients for the Dirlik method [73].

The cycle amplitude PDF is given by Dirlik:

$$p_{\mathrm{a,DK}}(s) = \frac{1}{\sqrt{m_0}} \left[\frac{G_1}{Q} \, \mathrm{e}^{\frac{-Z}{Q}} + \frac{G_2 \, Z}{R^2} \, \mathrm{e}^{\frac{-Z^2}{2R^2}} + G_3 \, Z \, \mathrm{e}^{\frac{-Z^2}{2}} \right], \tag{4.23}$$

where Z stands for the normalized amplitude:

$$Z = \frac{s}{\sqrt{m_0}}. \tag{4.24}$$

The remaining parameters G_1 to G_3, R, and Q are defined as [80]

$$G_1 = \frac{2\left(x_m - \alpha_2^2\right)}{1 + \alpha_2^2}, \qquad G_2 = \frac{1 - \alpha_2 - G_1 + G_1^2}{1 - R}, \tag{4.25}$$

$$G_3 = 1 - G_1 - G_2, \qquad R = \frac{\alpha_2 - x_m - G_1^2}{1 - \alpha_2 - G_1 + G_1^2}, \tag{4.26}$$

$$Q = \frac{1.25\,(\alpha_2 - G_3 - G_2\,R)}{G_1}, \qquad x_m = \frac{m_1}{m_0}\left(\frac{m_2}{m_4}\right)^{\frac{1}{2}}, \tag{4.27}$$

where x_m is the mean frequency, and α_2 is defined by Eq. (2.49). The closed-form expression that can be used to calculate the damage intensity of the random loading is of the form [80]

$$\begin{aligned} d^{\mathrm{DK}} = C^{-1}\,\nu_{\mathrm{p}}\,m_0^{\frac{k}{2}} \Bigg[& G_1\,Q^k\,\Gamma(1+k) \\ & + \left(\sqrt{2}\right)^k \Gamma\left(1 + \frac{k}{2}\right) \left(G_2\,|R|^k + G_3\right) \Bigg]. \end{aligned} \tag{4.28}$$

4.3.6 Zhao–Baker

Developed by Zhao and Baker [71] in 1992, this method gives the rainflow-cycle amplitude distribution in a form of a linear combination of Weibull and Rayleigh

PDFs:

$$p_{\mathrm{a}}(Z) = \overbrace{w\,\alpha\,\beta\,Z^{\beta-1}\,\mathrm{e}^{-\alpha\,Z^{\beta}}}^{\text{Weibull}} + \overbrace{(1-w)\,Z\,\mathrm{e}^{-\frac{Z^2}{2}}}^{\text{Rayleigh}}, \tag{4.29}$$

where Z is the normalized amplitude (4.24), also used in the Dirlik method, w is the weight coefficient defined as

$$w = \frac{1-\alpha_2}{1-\sqrt{\frac{2}{\pi}}\,\Gamma\left(1+\frac{1}{\beta}\right)\,\alpha^{-1/\beta}}, \tag{4.30}$$

and α and β are the Weibull distribution coefficients:

$$\alpha = 8 - 7\,\alpha_2, \tag{4.31}$$

$$\beta = \begin{cases} 1.1, & \alpha_2 < 0.9, \\ 1.1 + 9\,(\alpha_2 - 0.9), & \alpha_2 \geq 0.9. \end{cases} \tag{4.32}$$

The Rayleigh part (4.29) is largely responsible for high amplitudes, and the Weibull part has more impact at small amplitudes.

For the value $\alpha_2 < 0.13$, the coefficient w becomes greater than 1, which is an erroneous result for this specific spectral method. The Zhao–Baker method should therefore be avoided in case of processes with such an (uncommonly) low value of α_2. Furthermore, the results for materials with S-N curve coefficient $k > 6$ should be evaluated carefully, as the method was tuned in simulations with material parameters in the range of $2 \leq k \leq 6$.

A special case of the spectral method is derived for the case of S-N coefficient $k = 3$. Since the damage by means of rainflow depends more on $\alpha_{0.75}$ than on α_2 for low values of k, Zhao and Baker suggest an improved formulation for such case, where α is calculated as

$$\alpha = p^{-\beta}, \tag{4.33}$$

and p is the root of the equation [80]

$$\begin{aligned} \Gamma\left(1+\frac{3}{\beta}\right)(1-\alpha_2)\,p^3 + 3\,\Gamma\left(1+\frac{1}{\beta}\right)(\rho_{\mathrm{ZB}}\,\alpha_2 - 1)\,p + \\ +3\sqrt{\frac{\pi}{2}}\,\alpha_2\,(1-\rho_{\mathrm{ZB}}) = 0, \end{aligned} \tag{4.34}$$

with the correction factor ρ_{ZB} for $k = 3$ being

$$\rho_{\mathrm{ZB2}}|_{k=3} = \begin{cases} -0.4154 + 1.392\,\alpha_{0.75}, & \alpha_{0.75} \geq 0.5, \\ 0.28, & \alpha_{0.75} < 0.5. \end{cases} \tag{4.35}$$

Then we write the final expression for the Zhao–Baker method [71,80] as

$$d^{\mathrm{ZB1/ZB2}} = \frac{\nu_{\mathrm{p}}}{C}\, m_0^{k/2} \left[w\, \alpha^{-\frac{k}{\beta}}\, \Gamma\left(1+\frac{k}{\beta}\right) + (1-w)\, 2^{k/2}\, \Gamma\left(1+\frac{k}{2}\right) \right], \tag{4.36}$$

where ZB1 marks the original method when α and β are obtained using Eqs. (4.31)–(4.32), and ZB2 marks the improved formulation for the case of $k=3$ when α and p are obtained using Eqs. (4.33)–(4.35).

4.3.7 Tovo–Benasciutti

Rychlik [77] showed that the upper limit of the damage intensity for a Gaussian process is equal to the narrowband approximation d^{NB}, while Madsen et al. [81] gave an approximation to the lower limit of damage intensity d^{RC}, coincident with the range-count counting method in the time domain [67,72,81]:

$$d^{\mathrm{RC}} = \nu_{\mathrm{p}}\, C^{-1} \left(\sqrt{2 m_0}\, \alpha_2\right)^k \Gamma\left(1+\frac{k}{2}\right). \tag{4.37}$$

As the expected peak frequency ν_{p} (2.66) is related to the zero crossing frequency ν_0^+ (2.62), we have

$$\nu_{\mathrm{p}} = \frac{1}{2\pi}\sqrt{\frac{m_4}{m_2}} = \underbrace{\frac{1}{2\pi}\sqrt{\frac{m_2}{m_0}}}_{\nu_0^+} \underbrace{\sqrt{\frac{m_0\, m_4}{m_2^2}}}_{\alpha_2^{-1}}, \tag{4.38}$$

and the range-count damage estimator can be related to the narrowband damage estimator (4.10):

$$d^{\mathrm{RC}} = d^{\mathrm{NB}}\, \alpha_2^{k-1}. \tag{4.39}$$

In 2002, Tovo researched that the correct damage intensity estimate must lie somewhere between the lower d^{RC} and higher d^{NB} limits,

$$d^{\mathrm{RC}} \le d^{\mathrm{RFC}} \le d^{\mathrm{NB}}, \tag{4.40}$$

and proposed the following linear relationship [82]:

$$d^{\mathrm{TB}} = b\, d^{\mathrm{NB}} + (1-b)\, d^{\mathrm{RC}}. \tag{4.41}$$

Inserting (4.39) and reordering[2], we get

$$d^{\mathrm{TB}} = \left(b + (1-b)\,\alpha_2^{k-1}\right) d^{\mathrm{NB}}. \tag{4.42}$$

The d^{TB} stands for Tovo–Benasciutti as both researchers contributed to the development of this damage intensity estimation [82,72,67]. With regard to the definition of the missing weighting parameter b, two versions of the Tovo–Benasciutti method are known. Initially, b was approximated by Tovo as [82]

$$b_{\mathrm{app}}^{\mathrm{TB1}} = \min\left\{\frac{\alpha_1-\alpha_2}{1-\alpha_1}, 1\right\}. \tag{4.43}$$

The (4.43) is not continuous in α_1 and α_2, and in 2005, Benasciutti and Tovo [72] proposed another weighting estimation for the broadband random processes:

$$b_{\mathrm{app}}^{\mathrm{TB2}} = \frac{(\alpha_1-\alpha_2)\left[1.112\,(1+\alpha_1\alpha_2-(\alpha_1+\alpha_2))\,\mathrm{e}^{2.11\,\alpha_2}+(\alpha_1-\alpha_2)\right]}{(\alpha_2-1)^2}. \tag{4.44}$$

As it will be shown in the experimental part, (4.44) gives much more consistent and accurate results and was obtained using numerical simulations on a large group of spectra, shaped in a specific way, as shown in Fig. 4.6.

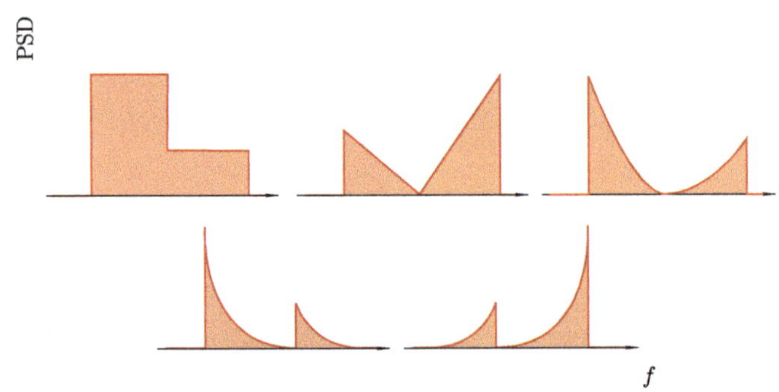

FIGURE 4.6 Shapes of spectra used in simulations to determine the coefficient b for the Tovo–Benasciutti method [60].

4.3.8 Other approaches

There are other approaches that were not explored in detail but might still be useful in certain cases. The bimodal spectral moments methods, formulated for

[2] It is worth noting that in several publications d^{TB} is wrongly defined as $\left(b+(1-b)\,\alpha_2^{k-1}\right)\alpha_2\, d^{\mathrm{NB}}$. All results in this book, and also the results in the authors' previous publications (e.g., [66,83]) were based on the computer code based on the correct definition (4.42).

analysis of random stress responses consisting of two modal peaks, constitute a specific category. Although not best suited for a general broadband random response, these methods can give good estimates if the response fits the assumptions. Some bimodal methods, available in the literature, are: the single-moment method by Lutes and Larsen [84,85], the Jiao–Moan method [86], the Sakai–Okamura method [87], and the Fu–Cebon method [88].

Another approach worth mentioning is the bands method by Braccesi et al. [89], which works by splitting the equivalent uniaxial stress PSD into narrow bands and then combining them according to a combination rule.

Chapter 5

Multiaxial vibration fatigue

Frequently, there arises a situation where the stress state in material is not simple *uniaxial* as considered in Ch. 4, but rather a more complex *multiaxial* one, which requires addressing the complete second-order stress tensor [90,91]. Good examples are: rotating shafts inside automotive components exposed to a combination of torsion and bending, pressurized vessels experiencing biaxial stresses even before potential additional loads are applied, offshore wind turbines exposed to random wave- and wind-induced loadings.

In the particularly important case of vibration fatigue, the structure is resonating due to vibration and deforms in specific (modal) shapes, which induce a multiaxial stress state in material. When multiple modes are jointly excited (by a broadband vibration load), the components of the stress are not in phase; it is said that the multiaxial stress state is nonproportional.

The uniaxial vibration fatigue analysis procedure is focused on cycle amplitudes and the assumption that each cycle causes a specific amount of damage, as dictated by the S-N curve. With a multiaxial stress state present in material, the effects of the interdependent stress components should be acknowledged. Specific rules must thus be applied, termed multiaxial criteria, to obtain a scalar stress function of time, which implicitly recognizes the combined effect of the second-order stress tensor components. This is a commonly adopted approach, which enables, by reduction of the loading to the equivalent uniaxial stress, application of spectral moments methods as presented in Chapter 4. The reduction can be made indiscriminately of the combined stress components effects, using an algorithm like the multiaxial rainflow counting [92], or, preferably, a multiaxial model can be applied that considers the properties of the physical fatigue process.

Quite a few models defined in this way are similar to the *plastic yield criteria*, namely the *von Mises* or *Tresca* approach. Others focus on the stresses on the critical plane, which is determined beforehand according to a specific criterion. Through these models, the experimentally confirmed effects, for example, of static tension or torsion, are acknowledged, and finally the time-to-failure estimate can be obtained [93,94].

Vibration Fatigue by Spectral Methods. https://doi.org/10.1016/B978-0-12-822190-7.00012-8

5.1 Time domain basics

The very straightforward ways of calculating the equivalent stress σ_{eq}, as inspired by the plastic yield criteria, are the maximum normal stress

$$\sigma_{eq} = \sigma_1, \tag{5.1}$$

the maximum shear stress

$$\tau = \frac{\sigma_{eq}}{2} = \frac{\sigma_1 - \sigma_3}{2}, \tag{5.2}$$

and the von Mises criterion

$$\sigma_{eq} = \frac{1}{\sqrt{2}} \sqrt{(\sigma_1 - \sigma_2)^2 + (\sigma_2 - \sigma_3)^2 + (\sigma_3 - \sigma_1)^2}, \tag{5.3}$$

where σ_1, σ_2, and σ_3 are the principal stresses (1.212). These are all able to deal with loads consisting of, for example, bending and torsion, but are only appropriate when the loading is *proportional*, meaning that the angles of the principal axes are fixed in time.

Exhaustive experimental and theoretical research was performed in this field, as early as 1950, by Gough and Pollard [95,96]. Parts of the results can be visualized by *Gough ellipses* shown in Fig. 5.1 [48]. They display the ratio of torsion and bending loads as they correspond to the *endurance limit*, the stress threshold below which no failure is thought to be predicted:

$$\frac{f^2}{\sigma_{af}^2} + \frac{q^2}{\tau_{af}^2} = 1 \qquad \text{(solid line)}, \tag{5.4}$$

$$\frac{q^2}{\tau_{af}^2} + \frac{f^2}{\sigma_{af}^2}\left(\frac{\sigma_{af}}{\tau_{af}} - 1\right) + \frac{f}{\sigma_{af}}\left(2 - \frac{\sigma_{af}}{\tau_{af}}\right) = 1 \qquad \text{(dashed line)}, \tag{5.5}$$

where f is the bending stress, σ_{af} is the bending fatigue limit, q is the shear stress, and τ_{af} is the torsion fatigue limit.

Comparing experimental results and estimates, given by the listed criteria, Sines [97] observed that it is hard to proclaim one of them as notably more appropriate than the others.

When the fatigue limit was studied for the pure *fully reversed bending and torsion* it was found, the ratio of bending and torsion fatigue limit σ_{af}/τ_{af} can differ significantly, depending on the chosen material, as pointed out in Table 5.1. Therefore using a single (fully reversed tension) S-N curve for an equivalent stress deduced for a multiaxial stress state is imprecise. Further, the σ_{af}/τ_{af} of bending/torsion fatigue limit can differ significantly from the chosen multiaxial criterion; see Table 5.2.

Complemented formulations were proposed based on empirical data. Both static torsion and bending effects were identified in the experimental data by

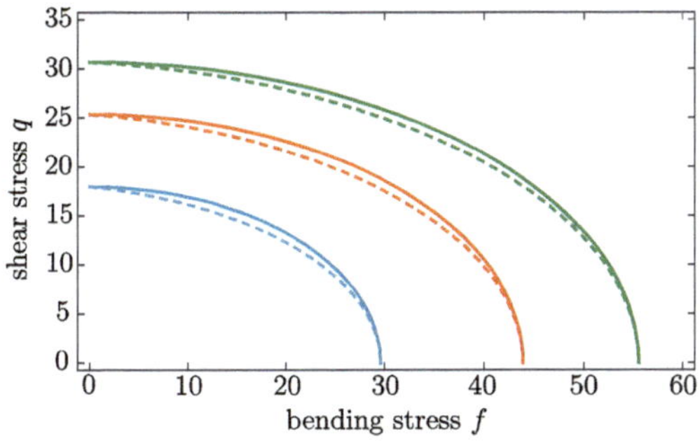

FIGURE 5.1 Gough ellipses for three different materials: full line (5.4) and dashed line (5.5). The area inside each ellipse is within the fatigue limit.

TABLE 5.1 Experimentally determined ratio σ_{af}/τ_{af} of bending/torsion fatigue limit [98].

Geometry	Material code	Material type	σ_{af}/τ_{af}
Plain	AlCuMg1	Aluminium alloy	1.66
Plain	C40 (SAE1040)	Carbon steel	1.35
Plain	CSN41 1523 (tip S355)	Structural steel	1.60
Plain	34CrMo4	Medium alloy steel	1.44
Plain	S20C (AISI 1020)	Low-alloy steel	2.32
Plain	D-30	Aluminium alloy	1.5
Plain	18G2A (S255 J0)	Structural steel	1.34
Plain	CuZn40Pb2	Brass	1.16
V-notched, $r = 0.5$ mm	C40 (SAE1040)	Carbon steel	0.77
Notched, $r = 5$ mm	Ck45 (SAE1045)	Carbon steel	1.58

TABLE 5.2 The σ_{af}/τ_{af} bending/torsion fatigue limit for different yield criteria [48].

	Max. shear stress	von Mises	Max. normal stress
σ_{af}/τ_{af}	2.0	1.73	1

Sines [97]. The effect of *static bending* is linear, higher stress meaning lower fatigue resistance. For static torsion, it was shown that the effect is negligible when it is below the plastic yield limit. Based on these findings, Sines [48,97] suggested a criterion of the form

$$\frac{\Delta\sigma_{eq}}{2} + \alpha\,(3\,\sigma_h) = \beta, \tag{5.6}$$

where $\Delta\sigma_{\mathrm{eq}}$ is obtained by the von Misses criterion (5.3), and σ_{h} is the *hydrostatic stress* defined using the first stress invariant (1.213):

$$\sigma_{\mathrm{h}} = \frac{I_1}{3}. \tag{5.7}$$

Sines' formulation requires relatively simple experiments to determine constants α and β and has the advantage to easily handle complex stress states.

Another formulation was suggested by Findley [99], based on experimentally supported assumption that the normal stress $\sigma_{\vec{\eta}}$ linearly affects the maximum allowed shear stress $\Delta\tau/2$ on the critical plane $\vec{\eta}$:

$$\left(\frac{\Delta\tau_{\vec{\eta}}}{2} + k\,\sigma_{\vec{\eta}}\right)_{\max} = f. \tag{5.8}$$

Sines and Findley formulations represent two different types of multiaxial criterion. Sines model from Eq. (5.6) is based on invariant of the stress tensor and gives a scalar value for damage in a material point. The Findley model from Eq. (5.8) gives damage on the *critical plane*, and the plane orientation needs to be determined in the process of fatigue analysis [100]. The research has shown [101–103] that the critical-plane type criteria are better suited for proportional loads (principal angles are fixed), whereas the invariant type criteria are better suited for *nonproportional loads* (the principal angles vary in time).

A subset of criteria that deserve mentioning are those that divide the structure in areas either above or below the material fatigue limit. Although this type of criteria can be used for durability assessment and usually also for calculating the location of the most critical point on the structure, they cannot be used directly to estimate the time-to-failure or damage intensity. Only by assessing the expected number of cycles to failure the S-N curve and Palmgren–Miner procedures can be applied. That is why here we consider the criteria that provide the estimate of the equivalent uniaxial stress only.

Just some of the formulations were described, and there are a number of conceptually different approaches to fatigue analysis that can be found in the literature [48,104,101,105,106,102]. In the following, we focus on those multiaxial criteria that provide a result in the form of the equivalent stress as a scalar function in time or frequency domain; however, for vibration fatigue, the final focus is the frequency-domain methods.

5.2 Random stress response

When performing fatigue analysis in the frequency domain, the stresses have to be defined in a certain probabilistic way, using the PSD instead of time history and assuming Gaussian probability of the random ergodic stress. The stress tensor is defined by the auto- and cross-spectral densities of its components (2.43).

In general, there are six independent components defining the stress tensor $\boldsymbol{\sigma}(t)$, which can be written in vector form (1.204):

$$\begin{aligned}\boldsymbol{\sigma}(t) = \mathbf{s}(t) &= \left[\sigma_{xx}(t), \sigma_{yy}(t), \sigma_{zz}(t), \sigma_{xy}(t), \sigma_{xz}(t), \sigma_{yz}(t)\right] \\ &= \left[s_1(t), s_2(t), s_3(t), s_4(t), s_5(t), s_6(t)\right],\end{aligned} \tag{5.9}$$

where s_1, s_2, and s_3 are the normal stresses, and s_4, s_5, and s_6 are the shear stresses.[1]

The frequency domain fatigue analysis routines also demand the information about the stress response in the frequency domain. As was shown in Eq. (1.222), the amplitude of the ith component of the stress steady-state response $\mathbf{X}_{s_i}(\omega)$ at N response locations is related to the N-dimensional excitation vector $\mathbf{F}(\omega)$ as follows:

$$\mathbf{X}_{s_i}(\omega) = \mathbf{H}_{s_i f}(\omega)\,\mathbf{F}(\omega), \tag{5.10}$$

where $\mathbf{H}_{s_i f}(\omega)$ is the stress frequency response function matrix from the excitation force to the response stress component s_i:

$$\mathbf{H}_{s_i f}(\omega) = \sum_{r=1}^{N} {}_r\mathbf{H}_{s_i f}(\omega) = \sum_{r=1}^{N} \frac{{}_r\mathbf{R}^{s_i}}{\omega_r^2 - \omega^2 + \mathrm{i}\,\eta_r\,\omega_r^2}. \tag{5.11}$$

Elements of the modal constant ${}_r\mathbf{R}^{s_i}$ are defined by Eq. (1.224), and the relation between the stress and displacement mode shape is defined by (1.221).

For the case of random process, the instantaneous combinations of components at times t are not explicitly known as in the time domain, they rather are given in a probabilistic way. They are characterized by the auto- (2.77) (PSD) and cross-correlation (2.78) (CSD) spectral density functions between different stress components.

Based on the transmissibility $\mathbf{H}_{s_i f}(\omega)$ (1.222) between the N-dimensional excitation force vector $\mathbf{F}(\omega)$ and the stress PSD (2.77), the component s_i of the N-dimensional PSD of the random response $\mathbf{S}_{s_i s_i}(\omega)$ is

$$\mathbf{S}_{s_i s_i}(\omega) = \left|\mathbf{H}_{s_i f}(\omega)\right|^2 \cdot \mathbf{S}_{ff}(\omega), \tag{5.12}$$

where $\mathbf{S}_{ff}(\omega)$ is the PSD of the force excitation $\mathbf{F}(\omega)$. A generalization of the PSD is the *stress spectral density matrix* (or *stress cross-correlation matrix*):

$$\mathbf{S}_{ss}(\omega) = \mathbf{H}_{sf}^{*}(\omega) \cdot \mathbf{S}_{ff}(\omega) \cdot \mathbf{H}_{sf}^{\mathrm{T}}(\omega), \tag{5.13}$$

where the *stress frequency response function vector* is defined as [15]

$$\mathbf{H}_{sf}(\omega) = \left[\mathbf{H}_{s_1 f}(\omega), \mathbf{H}_{s_2 f}(\omega), \mathbf{H}_{s_3 f}(\omega), \mathbf{H}_{s_4 f}(\omega), \mathbf{H}_{s_5 f}(\omega), \mathbf{H}_{s_6 f}(\omega)\right]^{\mathrm{T}}. \tag{5.14}$$

1 For clarity, we denote the stress by s.

The correlations between components $s_i(t)$ and $s_j(t)$ are the elements of the *cross-correlation matrix* $\mathbf{S}_{\mathbf{ss}}$, describing the random stress state in the frequency domain:

$$\mathbf{S}_{ss}(\omega) = \begin{bmatrix} \mathbf{S}_{s_1 s_1}(\omega) & \dots & \mathbf{S}_{s_1 s_6}(\omega) \\ \vdots & \ddots & \vdots \\ \mathbf{S}_{s_6 s_1}(\omega) & \dots & \mathbf{S}_{s_6 s_6}(\omega) \end{bmatrix}. \tag{5.15}$$

It makes sense to also define the covariance matrix $\mathbf{C}$, which plays an important role in fatigue analysis. For the case of the moment m_0 that corresponds to the covariance C_{ij},

$$m_{ij,0} = C_{ij} = \mathrm{Cov}\left\{s_i(t), s_j(t)\right\}. \tag{5.16}$$

Following Eq. (2.45) and then generalizing to all response nodes N

$$\mathbf{m}_{ij,0} = \mathbf{C}_{ij} = \int_{-\infty}^{+\infty} \mathbf{S}_{s_i s_j}^{\mathrm{T}}(\omega)\, \mathrm{d}\omega, \tag{5.17}$$

we write the *covariance matrix* $\mathbf{C}$ as

$$\mathbf{C} = \begin{bmatrix} \mathbf{C}_{11} & \dots & \mathbf{C}_{16} \\ \vdots & \ddots & \vdots \\ \mathbf{C}_{61} & \dots & \mathbf{C}_{66} \end{bmatrix}. \tag{5.18}$$

The covariance matrix gives the expected scatter of the respective stress tensor components and offers some insight when analyzing the multiaxial stress state.

5.3 Frequency-domain multiaxial criteria

We further present and experimentally compare a multitude of multiaxial criteria. They can be split into two groups, the first one with methods operating with the stresses on the critical plane and the second group with methods based on the stress tensor invariant, independent of the critical plane orientation. The studied criteria are formulated in the frequency domain and are suitable for use in the course of the vibration fatigue analysis procedure.

5.3.1 Critical plane criteria

Niesłony and Macha [107,108] generalized a group of critical plane multiaxial criteria, modeling the effect of combined shear and normal stress on the plane:

$$\max_t \left\{B\, \tau_{\vec{\eta}\,\vec{s}}(t) + K\, \sigma_{\vec{\eta}}(t)\right\} = F. \tag{5.19}$$

The damage depends on the combination of the stress $\sigma_{\vec{\eta}}(t)$ normal to the critical plane $\vec{\eta}$ and shear stress $\tau_{\vec{\eta}\,\vec{s}}$ in the direction of $\vec{s}$, as shown in Fig. 5.2; K and F are material constants, and B is determined by the distinct formulation of the multiaxial criterion.

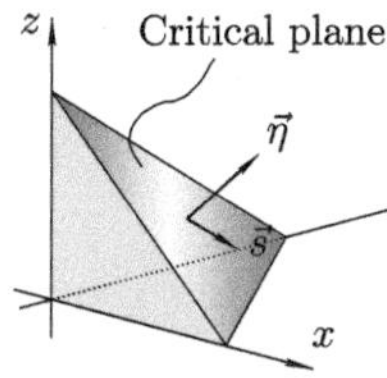

FIGURE 5.2 Vector of critical plane in normal $\vec{\eta}$ and tangential/shear direction $\vec{s}$.

The unit vector $\vec{\eta}$ is calculated by averaging the principal stress angles over time or by searching the plane that gives either the highest value or highest variance of the equivalent stress, depending on the chosen criterion.

The *equivalent stress* $\mathbf{S}_{\text{eq}}(\omega)$ in the frequency domain can be defined using one of the multiaxial criteria on the critical plane by a linear combination of the cross-correlation stress matrix $\mathbf{S}_{ss}(\omega)$ (5.15) with vector $\mathbf{a}$:

$$\mathbf{S}_{\text{eq}}(\omega) = \mathbf{a}\,\mathbf{S}_{ss}(\omega)\,\mathbf{a}^T, \tag{5.20}$$

where

$$\mathbf{a} = [a_1, a_2, a_3, a_4, a_5, a_6]. \tag{5.21}$$

Then three different criteria on the critical plane are given by the respective sets of the constants in Eq. (5.19). These are the *maximum normal stress*, *maximum shear stress*, and *maximum normal-and-shear stress* criteria. Each of them is defined by Eq. (5.20) and a suitable coefficient vector $\mathbf{a}$ as listed in Table 5.3. The l_i, m_i, and n_i are the mean directional cosines of the principal stress axes ($i = 1, 2, 3$; see Fig. 5.3 for $i = 1$), and K is the material parameter:

$$K = 2\,\frac{\tau_{\text{af}}}{\sigma_{\text{af}}} - 1, \tag{5.22}$$

where τ_{af} and σ_{af} are the fatigue limits in pure torsion and bending.

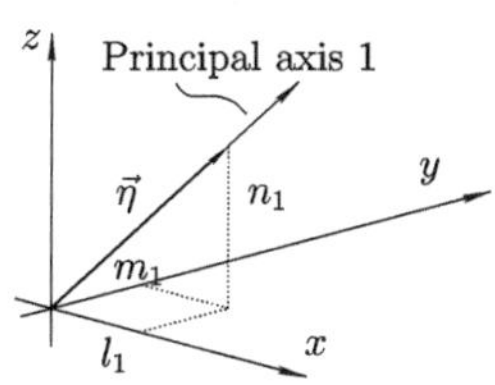

FIGURE 5.3 Directional cosines for the first principal axis: l_1, m_1, n_1.

TABLE 5.3 Coefficients defining the specific multiaxial stress criterion [108].

$\mathbf{a}_i$	Max. normal	Max. shear	Max. normal-and-shear
a_1	l_1^2	$\left(l_1^2 - l_3^2\right)$	$\frac{1}{1+K}\left[l_1^2 - l_3^2 + \kappa\left(l_1^2 + l_3^2\right)^2\right]$
a_2	m_1^2	$\left(m_1^2 - m_3^2\right)$	$\frac{1}{1+K}\left[m_1^2 - m_3^2 + K\left(m_1^2 + m_3^2\right)^2\right]$
a_3	n_1^2	$\left(n_1^2 - n_3^2\right)$	$\frac{1}{1+K}\left[n_1^2 - n_3^2 + K\left(n_1^2 + n_3^2\right)^2\right]$
a_4	$2l_1 m_1$	$2\,(l_1 m_1 - l_3 m_3)$	$\frac{1}{1+K}\left[l_1 m_1 - l_3 m_3 + K\,(l_1 + l_3)\,(m_1 + m_3)\right]$
a_5	$2l_1 n_1$	$2\,(l_1 n_1 - l_3 n_3)$	$\frac{1}{1+K}\left[l_1 n_1 - l_3 n_3 + K\,(l_1 + l_3)\,(n_1 + n_3)\right]$
a_6	$2m_1 n_1$	$2\,(m_1 n_1 - m_3 n_3)$	$\frac{1}{1+K}\left[m_1 n_1 - m_3 n_3 + K\,(m_1 + m_3)\,(n_1 + n_3)\right]$

Orientation of the critical plane

Because the resulting $\mathbf{S}_{\text{eq}}(\omega)$ heavily depends on the orientation of the critical plane (unit vector $\vec{\eta}$), the orientation must be chosen systematically and consistently across the numerical model. It is reasonable to assume that the crack will form on the plane where the damage according to the selected multiaxial criterion is the largest. A very straightforward but also pretty time-consuming approach to finding the maximum damage plane is calculating damages with the chosen criterion on many equally spaced planes. Then the one that gives maximum damage is chosen as critical. However, the plane search can be viewed as an optimization problem, where the maximum of damage intensity d as a function of plane rotations is searched. This way a broad selection of numerical optimization methods is available, and the result can be obtained much faster, yet still in a reliable way.

To further reduce the calculation time, another sensible assumption can be made: the variance of the equivalent stress $\text{Var}\left[\mathbf{S}_{\text{eq}}(\omega)\right]^{\vec{\eta}}$ on the plane $\vec{\eta}$ is proportional to the damage intensity $d^{\vec{\eta}}$. Then relying only on the variance to determine the plane orientation effectively reduces the input data from the frequency-dependent stress tensor $\mathbf{S}_{ss}(\omega)$ (5.15) only to the covariance matrix $\mathbf{C}$ (5.18). For a particular location, using the matrix $\mathbf{C}$, the plane is found that gives the largest variance $\sigma_{s_{\text{eq}}}^2$ for the chosen multiaxial criterion analogous to Eq. (5.20):

$$\sigma_{s_{\text{eq}}}^2 = \mathbf{a}\,\mathbf{C}\,\mathbf{a}^{\mathrm{T}}. \tag{5.23}$$

Lastly, there are special definitions to be used with some criteria. In particular, the definition by Matake [106] dictates that the critical plane is that with the maximum variance of the shear stress.

5.3.2 Equivalent von Mises stress (EVMS)

The equivalent von Mises stress (EVMS) criterion was suggested in 1994 by Preumont and Piéfort [109] and is based on the well-known von Mises plastic yield criterion, as defined in the time domain, where t is omitted for brevity:

$$s_c^2 = s_{xx}^2 + s_{yy}^2 + s_{zz}^2 - s_{xx}\, s_{yy} - s_{xx}\, s_{zz} - s_{yy}\, s_{zz} + 3\left(s_{xy}^2 + s_{yz}^2 + s_{xz}^2\right). \quad (5.24)$$

If the stress tensor is written in the form (5.9), then Eq. (5.24) can be formulated as a dot product:

$$s_c^2 = \mathbf{s}\,\mathbf{Q}\,\mathbf{s}^{\mathrm{T}} = \mathrm{Trace}\left\{\mathbf{Q}\,\mathbf{s}^{\mathrm{T}}\mathbf{s}\right\}, \quad (5.25)$$

where $\mathbf{Q}$ is the coefficient matrix

$$\mathbf{Q} = \begin{bmatrix} 1 & -1/2 & -1/2 & 0 & 0 & 0 \\ -1/2 & 1 & -1/2 & 0 & 0 & 0 \\ -1/2 & -1/2 & 1 & 0 & 0 & 0 \\ 0 & 0 & 0 & 3 & 0 & 0 \\ 0 & 0 & 0 & 0 & 3 & 0 \\ 0 & 0 & 0 & 0 & 0 & 3 \end{bmatrix}. \quad (5.26)$$

The expectation is

$$E\left[s_c^2\right] = \mathrm{Trace}\left\{\mathbf{Q}\,E\left[\mathbf{s}^{\mathrm{T}}\mathbf{s}\right]\right\}, \quad (5.27)$$

where the expected value $E\left[\mathbf{s}^{\mathrm{T}}\mathbf{s}\right]$ from Eq. (5.27) is in fact the mean-square of the stress components and can be obtained equivalently from the cross-spectral matrix $\mathbf{S}_{ss}(\omega)$ (5.15). Therefore the EVMS can be written in the frequency domain:

$$E\left[s_c^2\right] = \int_0^{\infty} \mathrm{Trace}\left\{\mathbf{Q}\,\mathbf{S}_{ss}(\omega)\right\} \mathrm{d}\omega. \quad (5.28)$$

Due to the equivalence of Eqs. (5.27) and (5.28), it follows that the root-mean-square (*RMS*) of the von Mises criterion is equal in the time and frequency domain. This forms the base assumption of EVMS, which can now be written as

$$\mathbf{S}_c(\omega) = \mathrm{Trace}\left\{\mathbf{Q}\,\mathbf{S}_{ss}(\omega)\right\}. \quad (5.29)$$

It is evident that the frequency-domain von Mises formulation is not fully equivalent to its time-domain counterpart. The RMS equality condition is very broad and allows for infinite solutions in the frequency domain that fulfill it. Nevertheless, the EVMS is a very useful and widely used method.

There are some weaknesses though, which have been pointed out by researchers. Benasciutti [98] compared the accuracy of the EVMS for a case of pure normal stress and pure shear stress, showing that the results (accuracy) are largely dependent on the material parameters. The estimates match only when the ratio of fatigue limits is $\sigma_{\mathrm{af}} = \sqrt{3}\,\tau_{\mathrm{af}}$ and the slope coefficients are equal for normal and shear stress S-N curves.

Another fact to note is that EVMS formulation disregards the phase angles between the stress components. This was addressed by Bonte et al. [110], who proposed a formulation that accounts for phase angles and showed that the EVMS overestimates the maximum von Mises stress for out-of-phase biaxial stresses. However, EVMS also shows a better fit to experimental data compared to the formula by Bonte et al., as pointed out by Niesłony [107].

5.3.3 Carpinteri–Spagnoli (C-S)

The C-S criterion has originally been suggested in 2001 by Carpinteri and Spagnoli [111] as a time-domain formulation, which determines the critical plane by averaging the Euler angles of the principal stress axes over time. Recently, in 2014, it has been adapted for use in the frequency domain, where the critical plane is deduced by criterion of maximum stress rather than by principal directions averaging [112,113].

The cross-spectral matrix $\mathbf{S}_{ss}(\omega)$ is transformed using the matrix $\mathbf{C}_{cs}$ (not to be confused with the covariance matrix $\mathbf{C}$) according to

$$\mathbf{S}'_{ss}(\omega) = \mathbf{C}_{cs}\,\mathbf{S}_{ss}(\omega)\,\mathbf{C}_{cs}^{\mathrm{T}}, \tag{5.30}$$

where $\mathbf{C}_{cs}$ is defined by the angles $c_\psi = \cos\psi$, $s_\psi = \sin\psi$, $c_\theta = \cos\theta$, $s_\theta = \sin\theta$, $c_\phi = \cos\phi$, and $s_\phi = \sin\phi$:

$$\mathbf{C}_{cs} = \begin{bmatrix} 1 & 0 & 0 & 0 & 0 & 0 \\ 0 & c_\psi^2 & s_\psi^2 & 2\,c_\psi s_\psi & 0 & 0 \\ 0 & s_\psi^2 & c_\psi^2 & -2\,s_\psi c_\psi & 0 & 0 \\ 0 & -2\,c_\psi s_\psi & 2\,c_\psi s_\psi & c_\psi^2 - s_\psi^2 & 0 & 0 \\ 0 & 0 & 0 & 0 & c_\psi & -s_\psi \\ 0 & 0 & 0 & 0 & s_\psi & c_\psi \end{bmatrix} \cdot \begin{bmatrix} c_\theta^2 & 0 & s_\theta^2 & 0 & 2\,c_\theta s_\theta & 0 \\ 0 & 1 & 0 & 0 & 0 & 0 \\ s_\theta^2 & 0 & c_\theta^2 & 0 & -2\,c_\theta s_\theta & 0 \\ 0 & 0 & 0 & c_\theta & 0 & -s_\theta \\ -c_\theta s_\theta & 0 & c_\theta s_\theta & 0 & c_\theta^2 - s_\theta^2 & 0 \\ 0 & 0 & 0 & s_\theta & 0 & c_\theta \end{bmatrix} \cdot \tag{5.31}$$

$$\cdot\begin{bmatrix} 1 & 0 & 0 & 0 & 0 & 0 \\ 0 & c_\phi^2 & s_\phi^2 & 2\,c_\phi s_\phi & 0 & 0 \\ 0 & s_\phi^2 & c_\phi^2 & -2\,s_\phi c_\phi & 0 & 0 \\ 0 & -2\,c_\phi s_\phi & 2\,c_\phi s_\phi & c_\phi^2 - s_\phi^2 & 0 & 0 \\ 0 & 0 & 0 & 0 & c_\phi & -s_\phi \\ 0 & 0 & 0 & 0 & s_\phi & c_\phi \end{bmatrix}.$$

By varying the angles ϕ and θ ($0 \le \phi \le 2\pi$ and $0 \le \theta \le \pi$) the expected value of the extreme of the random process s'_z is found [114]:

$$E\left[\max s'_z\right] = \sqrt{m_0}\sqrt{2\,\ln N_1} + \frac{0.5772}{\sqrt{2\,\ln N_1}}, \tag{5.32}$$

where N_1 is the expected number of cycles calculated from the duration of the loading T as $N_1 = \nu_0^+ T$. In the next step the maximum m_0 for $S'_{66}(\omega)$ is found by varying ψ.

Lastly, the normal to the critical plane is calculated by determining the offset angle δ, dependent on the ratio of fatigue limits in pure axial and shear stress, σ_{af} and τ_{af}:

$$\delta = \frac{3\pi}{8}\left[1 - \left(\frac{\tau_{\mathrm{af}}}{\sigma_{\mathrm{af}}}\right)^2\right]. \tag{5.33}$$

With the critical plane determined, the equivalent stress $\mathbf{S}_{\mathrm{eq}}(\omega)$ is calculated through a linear combination of the normal stress and shear stress projected on the critical plane:

$$\mathbf{S}_{\mathrm{eq}}(\omega) = \mathbf{S}_{s_3 s_3}(\omega) + \left(\frac{\sigma_{\mathrm{fr}}}{\tau_{\mathrm{fr}}}\right)\mathbf{S}_{s_2 s_3}(\omega). \tag{5.34}$$

The components $\mathbf{S}_{s_3 s_3}(\omega) = \mathbf{S}_{zz}(\omega)$ and $\mathbf{S}_{s_2 s_3}(\omega) = \mathbf{S}_{yz}(\omega)$ are defined so that x and y determine the critical plane and are perpendicular to z.

5.3.4 Projection-by-Projection (PbP)

The PbP approach was developed in 2011 by Cristofori et al. [115] and is a reformulation of the time-domain approach, first proposed by Cristofori et al. [116] in 2008.

The stress invariant $\mathbf{s}_{\mathrm{inv}} = [S_1, S_2, S_3, S_4, S_5]$ is defined in the Euclidean space E_5:

$$S_1 = \frac{\sqrt{3}}{2}\,s'_1, \qquad S_2 = \frac{1}{2}\left(s'_2 - s'_3\right), \qquad S_3 = s'_4, \tag{5.35}$$

$$S_4 = s'_5, \qquad S_5 = s'_6, \tag{5.36}$$

where $\mathbf{s}' = [s'_1, s'_2, s'_3, s'_4, s'_5, s'_6]$ are the components of the stress tensor deviator, defined together with hydrostatic pressure s_h:

$$\mathbf{s} = \mathbf{s}' + \mathbf{I}\, s_h, \tag{5.37}$$

where $\mathbf{I}$ is the identity matrix. The cross-spectral matrix $\mathbf{S}_{\text{inv}} = \mathbf{s}_{\text{inv}}\, \mathbf{s}_{\text{inv}}^{\text{T}}$ is transformed into the subspace of the eigenvectors of the (invariant) covariance matrix $\mathbf{C}'$. First, the eigenvectors $\mathbf{U}$ of matrix $\mathbf{C}'$ are calculated, which transform $\mathbf{C}'$ into the diagonal matrix $\mathbf{C}'_0$:

$$\mathbf{C}'_0 = \mathbf{U}^T\, \mathbf{C}'\, \mathbf{U}. \tag{5.38}$$

Using the eigenvectors $\mathbf{U}$, we now transform the cross-spectral matrix $\mathbf{S}_{\text{inv}}(\omega)$ to $\mathbf{S}_U(\omega)$:

$$\mathbf{S}_U(\omega) = \mathbf{U}\, \mathbf{S}_{\text{inv}}\, \mathbf{U}, \tag{5.39}$$

which gives a group of five uncorrelated uniaxial processes $\Omega_i(\omega)$, the elements of the diagonal of $\mathbf{S}_U(\omega)$. Each of them is assigned the damage intensity $d\,(\Omega_i)$ using a spectral moments method. The damage intensities are then summed according to the quadratic rule:

$$d(\Omega) = \left[\sum_{i=1}^{5} \left(d(\Omega_i)\right)^{\frac{2}{k_{\text{ref}}}} \right]^{\frac{k_{\text{ref}}}{2}}. \tag{5.40}$$

The material coefficients $J_{A,\text{ref}}$ and k_{ref} are based on a reference S-N curve, which takes into account the ratio $\sigma_{\text{af}}/\tau_{\text{af}}$. The reference S-N curve is obtained by a combination of the pure axial ($J_{A,\text{axi}}$, k_{axi}) and torsion ($J_{A,\text{tor}}$, k_{tor}) definitions:

$$\begin{aligned} J_{A,\text{ref}} &= J_{A,\text{tor}} + \rho_{\text{ref}} \left(J_{A,\text{axi}} - J_{A,\text{tor}}\right), \\ k_{\text{ref}} &= k_{\text{tor}} + \rho_{\text{ref}} \left(k_{\text{axi}} - k_{\text{tor}}\right). \end{aligned} \tag{5.41}$$

The ρ_{ref} coefficient describes the multiaxiality of the load in material and is defined as

$$\rho_{\text{ref}} = \sqrt{3}\, \frac{s_{\text{h,m}} + \sqrt{2\, m_{0,h}}}{\sqrt{2 \sum_{i=1}^{5} m_{0,i}}}, \tag{5.42}$$

where $s_{h,m}$ is the mean hydrostatic pressure, and $m_{0,h}$ and $m_{0,i}$ are the zeroth spectral moments of the hydrostatic pressure and the ith uncorrelated uniaxial process $\Omega_i(\omega)$.

Chapter 6

Vibration fatigue analysis using modal decomposition

The vibration fatigue analysis is performed to estimate the time-to-failure based on the steady-state dynamic response of a structure. Because of that fact, the modal reduction and modal superposition approaches can be used to great advantage. By sacrificing some accuracy the analysis procedure can be made considerably more numerically efficient (and therefore faster) in comparison with the more conventional time-integration approaches. Further, a big benefit is gained in an exposed explicit relation between the dynamic properties of the structure and the damage intensity, which can then be evaluated for design improvements. Fast structural optimization for fatigue, generating many iterations on design, thus becomes feasible.

Modal reduction is a procedure often used in the structural dynamics analysis field and has already been extensively investigated in Sec. 1.2.6. It is used to reduce the complete modal model, containing every modal shape, to the incomplete model, comprising only a subset of those shapes. The subset is picked taking into account the excitation frequency range and other case-specific characteristics. There are studies available in the literature that apply the fatigue analysis to the modal parameters of the structural model [117–120], but they do not, with very few exceptions [120], take the full advantage offered by the reduction procedure.

We present a new approach to vibration fatigue analysis using modal decomposition, which has been first presented by Mršnik et al. [74]. The damage intensity d in material is determined from the dynamic properties of the structure, and the direct connection between d and the excitation force PSD $\mathbf{S}_{ff}(\omega)$ is given through a function g of modal parameters $\mathbf{\Phi}_r, \omega_r, \eta_r$ of a linear dynamic system:

$$d = g\left(\mathbf{S}_{ff}(\omega), \mathbf{\Phi}_r, \omega_r, \eta_r\right), \tag{6.1}$$

where the modal parameters are the modal shape $\mathbf{\Phi}_r$, modal frequency ω_r, and damping loss factor η_r.

Through the implementation of the modal decomposition procedure, the analysis itself is simplified. It also offers means of straightforward application on experimental strains obtained, for instance, in experimental strain modal analysis.

Vibration Fatigue by Spectral Methods. https://doi.org/10.1016/B978-0-12-822190-7.00013-X

6.1 Reduced modal stress response

Considering a standard finite element model, the total number of modes is equal to the (often very large) number of DOF, not all of which are equally relevant for the stress response solution. Standard accelerated vibration test profiles, for example, tend to be limited to the frequency range of up to 2000 Hz. The contribution of high-frequency modes is negligible, and they can be truncated to a subset of m modes while sacrificing hardly any accuracy. The reduced modal model is based on Eq. (5.11), whereas instead of all available mode shapes N, we consider only a subset of m modes:

$$\widetilde{\mathbf{H}}_{s_i f}(\omega) = \sum_{r=1}^{m<N} {}_r\mathbf{H}_{s_i f}(\omega). \tag{6.2}$$

It follows that the reduced stress cross-correlation matrix (5.13) is

$$\widetilde{\mathbf{S}}_{ss}(\omega) = \widetilde{\mathbf{H}}^*_{sf}(\omega) \cdot \mathbf{S}_{ff}(\omega) \cdot \widetilde{\mathbf{H}}^{\mathrm{T}}_{sf}(\omega), \tag{6.3}$$

where the reduced stress frequency response function vector is (5.14)

$$\widetilde{\mathbf{H}}_{sf}(\omega) = [\widetilde{\mathbf{H}}_{s_1 f}(\omega), \widetilde{\mathbf{H}}_{s_2 f}(\omega), \widetilde{\mathbf{H}}_{s_3 f}(\omega), \widetilde{\mathbf{H}}_{s_4 f}(\omega), \widetilde{\mathbf{H}}_{s_5 f}(\omega), \widetilde{\mathbf{H}}_{s_6 f}(\omega)]^{\mathrm{T}}. \tag{6.4}$$

Next, we make approximation, crucial to the formulation of the modal decomposition approach. The assumption that the modes participating in response are well separated may justify approximating the expression from Eq. (6.3) by factoring out the sum, because the single mode should only contribute significantly in the close vicinity of its natural frequency:

$$\widetilde{\mathbf{H}}^*_{sf}(\omega) \cdot \mathbf{S}_{ff}(\omega) \cdot \widetilde{\mathbf{H}}^{\mathrm{T}}_{sf}(\omega) \approx \sum_{r=1}^{m<N} {}_r\mathbf{H}^*_{sf}(\omega) \cdot \mathbf{S}_{ff}(\omega) \cdot {}_r\mathbf{H}^{\mathrm{T}}_{sf}(\omega). \tag{6.5}$$

By (5.13) and (5.14) the elements of the approximated reduced stress cross-correlation matrix (6.3) are

$$\widetilde{\mathbf{S}}'_{s_i s_j}(\omega) = \sum_{r=1}^{m<N} \frac{{}_r\mathbf{R}^*_{s_i} \cdot \mathbf{S}_{ff}(\omega) \cdot {}_r\mathbf{R}^{\mathrm{T}}_{s_j}}{\omega^4 - 2\,\omega^2\,\omega_r^2 + \left(1+\eta_r^2\right)\omega_r^4}, \tag{6.6}$$

where (1.217) defines the matrix ${}_r\mathbf{R}_{s_j}$. Using

$${}_r\mathbf{R}_s = [{}_r\mathbf{R}_{s_1}, {}_r\mathbf{R}_{s_2}, {}_r\mathbf{R}_{s_3}, {}_r\mathbf{R}_{s_4}, {}_r\mathbf{R}_{s_5}, {}_r\mathbf{R}_{s_6}], \tag{6.7}$$

we can write the approximated reduced stress cross-correlation matrix

$$\widetilde{\mathbf{S}}'_{ss}(\omega) = \begin{bmatrix} \widetilde{\mathbf{S}}'_{s_1 s_1}(\omega) & \dots & \widetilde{\mathbf{S}}'_{s_1 s_6}(\omega) \\ \vdots & \ddots & \vdots \\ \widetilde{\mathbf{S}}'_{s_6 s_1}(\omega) & \dots & \widetilde{\mathbf{S}}'_{s_6 s_6}(\omega) \end{bmatrix} \tag{6.8}$$

as

$$\widetilde{\mathbf{S}}'_{ss}(\omega) = \sum_{r=1}^{m<N} \zeta_r(\omega) \cdot {}_r\mathbf{R}_s^* \cdot \mathbf{S}_{ff}(\omega) \cdot {}_r\mathbf{R}_s^{\mathrm{T}} = \sum_{r=1}^{m<N} {}_r\mathbf{S}'_{ss}(\omega). \tag{6.9}$$

The dynamic part is characterized by $\zeta_r(\omega)$, and ${}_r\mathbf{S}'_{s_i s_j}(\omega)$ is the stress response PSD pertaining to the rth mode.

6.1.1 Equivalent uniaxial stress

The Preumont and Piéfort [109] frequency domain formulation of the equivalent von Mises stress (EVMS) can be applied straightforwardly on the sum of modal stress responses $\widetilde{\mathbf{S}}'_{ss}(\omega)$ from Eq. (6.8). Based on the definition of EVMS from Section 5.3.2, the approximated equivalent uniaxial stress $\mathbf{S}'_c(\omega)$ is then defined as

$$\widetilde{\mathbf{S}}'_c(\omega) = \sum_{r=1}^{m<N} {}_r\mathbf{S}'_c(\omega) = \sum_{r=1}^{m<N} \mathrm{Trace}\left[\mathbf{Q} \cdot {}_r\widetilde{\mathbf{S}}'_{ss}(\omega)\right], \tag{6.10}$$

where ${}_r\widetilde{\mathbf{S}}'_{ss}(\omega)$ relates to the rth mode of $\widetilde{\mathbf{S}}'_{ss}(\omega)$; see Eqs. (6.8) and (6.9). Although such an approach is similar to that proposed by Braccesi et al. [120], it takes a step further using approximation to retain the modal properties throughout the complete analysis procedure.

6.2 Damage intensity

A spectral moments method of choice can be used on the equivalent uniaxial stress $\widetilde{\mathbf{S}}'_c(\omega)$. This requires the spectral moments to be calculated first, and due to the associativity of integral sums, the spectral moment vector (at response degrees of freedom) of order i for the reduced mode r can be calculated:

$${}_r\mathbf{m}_i = 2 \int_0^{\infty} \omega^i \cdot {}_r\mathbf{S}_c(\omega)\, \mathrm{d}\omega, \tag{6.11}$$

and after substitution into the previous expressions, we get

$${}_r\mathbf{m}_i = 2 \cdot \mathrm{Trace}\left[\mathbf{Q} \cdot {}_r\mathbf{R}_s^* \cdot \int_0^{\infty} \mathbf{S}_{ff}(\omega) \cdot \omega^i\, \zeta_r(\omega)\, \mathrm{d}\omega \cdot {}_r\mathbf{R}_s^{\mathrm{T}}\right]. \tag{6.12}$$

The expressions derived so far make it possible to obtain the ith spectral moment for the specific mode r while calculating the value of the integral in spectral moment definition only once for each mode. Without taking the assumptions and approximating in Eq. (6.5), the spectral moments would need evaluating in each material node of the numerical model. Reducing the number of DOF and consequently also the number of modal shapes, participating in response, the calculation of solution can thus be accelerated by multiple orders of magnitude.

Once all the required spectral moment vectors ${}_r\mathbf{m}_i$ are obtained, they are summed to produce the total moment $\tilde{\mathbf{m}}_i$ for the reduced modal model:

$$\tilde{\mathbf{m}}_i = \sum_{r=1}^{m<N} {}_r\mathbf{m}_i. \tag{6.13}$$

The procedure that then follows is similar to the standard damage intensity calculation using spectral moment methods. The Tovo–Benasciutti method of calculating damage intensity d^{TB} gave good results in the comparison study (4.42); for details, see Section 4.3.7.

6.2.1 Modal damage contributions

One substantial benefit of using the described approach to evaluation of vibration fatigue life is that the damage intensity can be attributed directly to modal properties of the system. A measure of damage intensity D_r is devised, which can characterize the portion of the damage intensity contributed by the respective mode.

The resonances contributing to damage cannot be combined in a straightforward way due to how the damaging cycles are formed, as described in Chapters 4 and 5. The partial damage intensity $d_{\setminus M_r}$ is first calculated, which accounts for the total damage intensity but with excluded mode r. The $d_{\setminus M_r}$ is then subtracted from the total damage intensity by all participating modes d. The procedure is repeated for each mode in the subset of the incomplete model, resulting in a spectrum type of data, showing the severity of each mode as compared to others:

$$D_r = \frac{d - d_{\setminus M_r}}{d} = 1 - \frac{d_{\setminus M_r}}{d}, \tag{6.14}$$

which can be calculated very efficiently, once the total damage intensity d has been obtained, as all the partial solutions (modal spectral moments ${}_r\mathbf{m}_i$ (6.13)) have already been determined.

It is worth noting that the critical location estimate can change when one of the modes is excluded. This can offer additional insight into the problematic locations on the structure. However, usually the critical location will be fixed to the critical location of total damage intensity d while calculating the measure D_r.

Using a similar procedure, other measures can be devised, which characterize the effects of specific properties of the linear dynamic system, such as damping, modal constants, or excitation.

6.3 Numerical experiment

The application and performance of the vibration fatigue analysis using modal decomposition procedure are demonstrated on a clamped beam, shown in Fig. 6.1, which is excited by random vibrations.

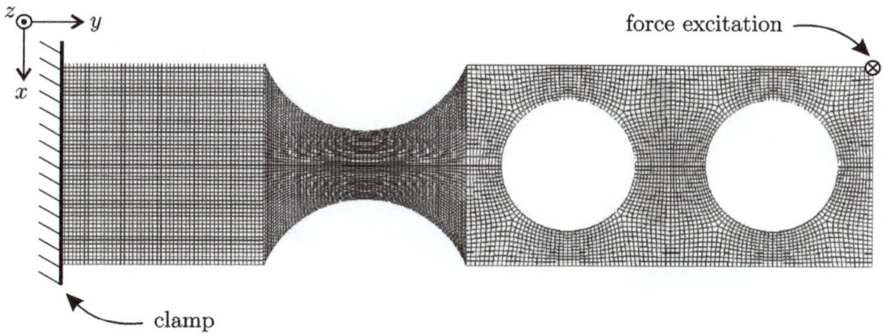

FIGURE 6.1 FE model of the cantilever beam.

A numerical model of the beam is prepared using the finite element method (FEM). The material is 0.5-mm-thick steel, which is fixed at one end. The broadband random excitation profile $\mathbf{S}_{ff,z}$ is applied eccentrically on the free end, along the z-axis. The vibration profile is unit uniform and spans such a frequency range that it excites the first five modes $r = 1, 2, \ldots, 5$ of the linear model. The eigensolution is obtained for an undamped model; natural frequencies are listed in Table 6.1 together with the damping ratios applied in the modal superposition procedure. The displacement modal shapes are given in Figs. 6.2–6.6.

TABLE 6.1 Natural frequencies and damping of the clamped beam.

	Mode I	Mode II	Mode III	Mode IV	Mode V
ω_r [s^{-1}]	70	405	451	1064	1746
η_r [/]	0.025	0.031	0.028	0.021	0.034

For this particular numerical example, where the forced excitation is uniform and applied only along the z-axis, the expression from Eq. (6.12) can be simplified to

$$
{}_r\mathbf{m}_i = \underbrace{2 \cdot \int_0^\infty \omega^i \, \zeta(\omega) \, \mathrm{d}\omega}_{{}_rI_i} \cdot \mathrm{Trace}\left[\mathbf{Q} \cdot {}_r\mathbf{R}_s^* \cdot \mathbf{S}_{ff,z} \cdot {}_r\mathbf{A}_s^{\mathrm{T}}\right], \tag{6.15}
$$

FIGURE 6.2 Displacement modal shape I.

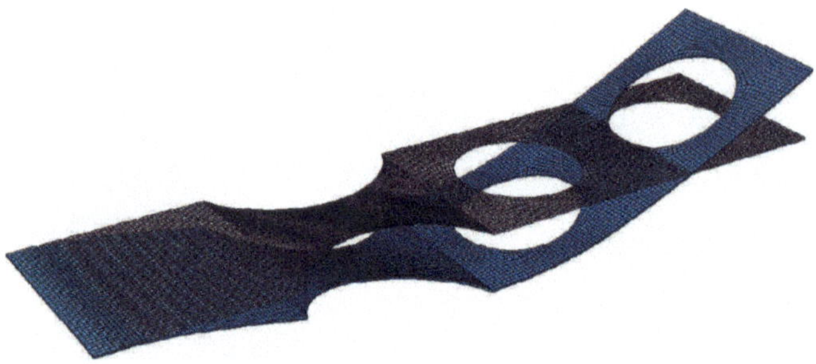

FIGURE 6.3 Displacement modal shape II.

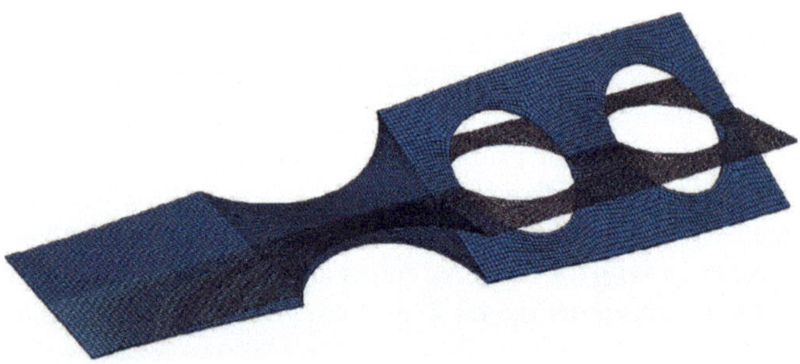

FIGURE 6.4 Displacement modal shape III.

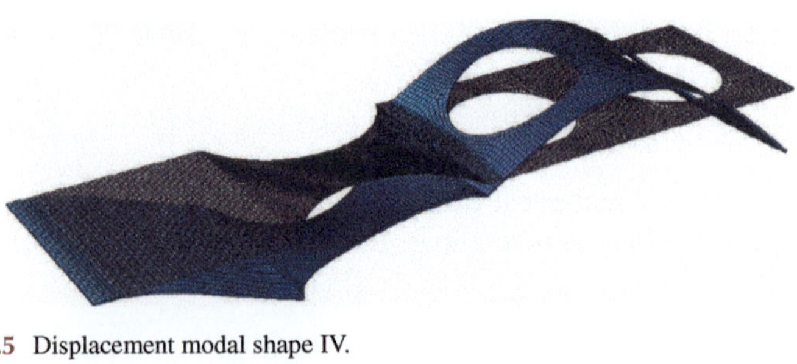

FIGURE 6.5 Displacement modal shape IV.

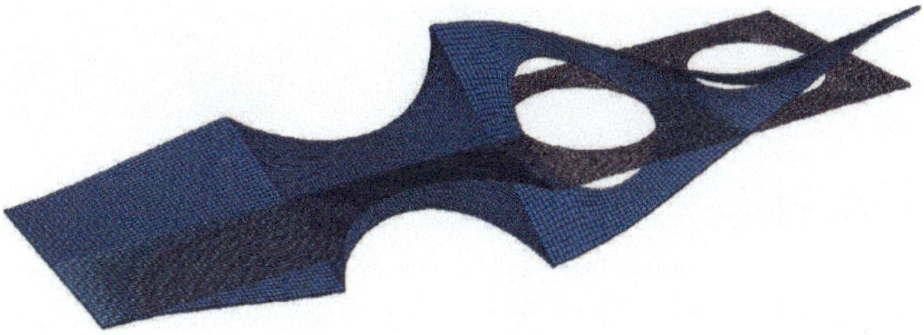

FIGURE 6.6 Displacement modal shape V.

where the integral ${}_rI_i$ is calculated once for a particular mode r. The results of these calculations are provided in Table 6.2, where the definite integrals have the upper bound of $\omega_{\max} = 5000$ rad/s. The Tovo–Benasciutti method is used to obtain the final total damage intensity d_{TB} with S-N curve parameters $k = 3$ and $C = 10^{20}$. The results can be seen in Fig. 6.7.

TABLE 6.2 Integrals ${}_rI_i$ calculated at each mode (I–V) for each order of spectral moment (0, 1, 2, and 4).

	Mode I	Mode II	Mode III	Mode IV	Mode V	
${}_rI_0$	36.864	0.152	0.122	0.012	0.002	$\times 10^{-5}$
${}_rI_1$	25.55	0.612	0.546	0.131	0.030	$\times 10^{-3}$
${}_rI_2$	0.800	0.125	0.124	0.070	0.026	$\times 10^{+0}$
${}_rI_4$	1.877	5.090	6.043	16.820	16.87	$\times 10^{+4}$

A more standard procedure was performed, incorporating modal reduction but not the approximation from Eq. (6.5). The results, with regards to the damage intensity distribution on the numerical model, gave the exact same critical location, as seen in Fig. 6.8. However, the damage intensity estimate in this critical point shows that the new approach has a conservative relative error of 65% in comparison with the standard procedure. This is an acceptable error, considering the scatter of up to ±300%, which is often observed in vibration fatigue analysis. Fig. 6.9 shows the modal damage contributions (6.14): for the studied excitation conditions, the majority of damage comes from mode shape 1, followed by the mode shape 3.

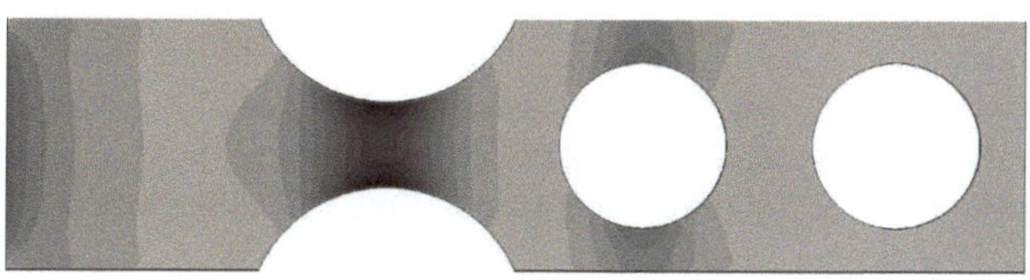

FIGURE 6.7 The calculated damage intensity d for the nodes using the proposed approach (with approximation). Darker shade indicates higher d.

FIGURE 6.8 The calculated damage intensity d for the nodes using the classical approach (without the approximation). Darker shade indicates higher d.

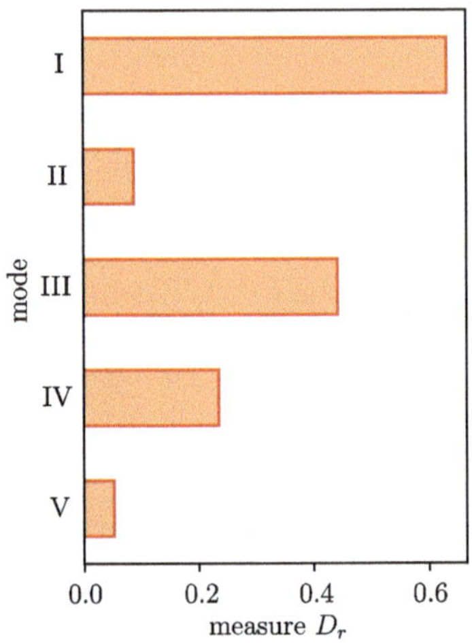

FIGURE 6.9 The comparison of damage contributions attributed to each participating modal shape.

6.4 Discussion

The evaluated numerical model comprises 18 569 nodes, and 5×4 evaluations of the spectral moment integral were needed, as opposed to the $18\,569 \times 4$ evaluations, which would have to be performed using the conventional approach, integrating for each node. This makes the new approach on the order of 10^3 faster for the specific example considered here. The speed advantage will, however, increase proportionally to the number of DOF in relation to the number of DOF of the incomplete model.

Because the spectral moment integral is defined relative to dynamic properties of the system, it is possible to calculate the integrals very precisely. A dense frequency sampling is usually required in the vicinity of the resonance, where the amplitude rises sharply. For the case of the proposed numerical example, this integral can be evaluated symbolically in advance, thus achieving exceptional accuracy and, at the same time, saving time of calculation.

Chapter 7

Identification of nonstationary and non-Gaussian excitation

In vibration fatigue, flexible structures operate at or close to their natural frequencies. Therefore it is common to consider the input excitation and the stress/strain response of the structure to be Gaussian and stationary, and the assumptions discussed in Section 2.2 apply.

In a real loading cases, however, we often experience non-Gaussian excitation, which can lead to the non-Gaussian response and induces a shorter fatigue life. For this reason, non-Gaussianity has recently been attracting a lot of scientific focus [41,67,121–127]. Some examples of non-Gaussian loads include rail irregularities in railways, pressure fluctuations in the aerospace sector, and military environmental standards [128,129]. Rizzi et al. [130], Kihm et al. [131], and Niesłony et al. [132] investigated how non-Gaussian excitation affects the fatigue life of linear and nonlinear structures. Cianetti et al. [127] investigated the possibility of correcting the damage identified with the Gaussian-loads hypothesis by adopting a corrective coefficient they obtained as a function of non-Gaussian parameters. It has been numerically demonstrated that for a dynamic linear system operating close to the natural frequency, the non-Gaussian excitation results in a Gaussian response when the period of the system's impulse response is much greater than the rate of the peaks in the loading. Instead, for a nonlinear system, the response to Gaussian or non-Gaussian excitations was always non-Gaussian due to the violation of the central limit theorem [15]. Rizzi and Kihm obtained results for the case of kinematically excited structures.

Also, for real applications, the excitation is frequently nonstationary (e.g., a sudden load increase, changes in road roughness, turbulence loads) [133,134]. As discussed in Section 2.2.2, random processes are considered to be *weakly stationary* if the mean values and covariance functions are time independent and *strongly* (or strictly) stationary if the probability distributions are time independent. Whereas the definition of a nonstationary signal is clear, the *identification* and quantification of a nonstationarity signal is not. To identify the nonstationarity of an excitation, Rouillard [135] adopted the nonparametric "run-test" method for time-history signal evaluation to obtain the nonstationarity level of vehicle vibrations.

In this section, we give the main properties of non-Gaussian and nonstationary signals. The experimental evaluation of these signal types on the vibration fatigue life is presented later in Part II, Chapter 12.

Vibration Fatigue by Spectral Methods. https://doi.org/10.1016/B978-0-12-822190-7.00014-1

7.1 Identifying non-Gaussian signal

A random variable x is said to be Gaussian if its probability density function $p(x)$ is given by Eq. (2.24), where μ is the mean value, and σ is the standard deviation, which are correlated to the PDF as the first and second central moments (Eqs. (2.25) and (2.32)). In terms of discrete time-series data, the mean μ and the jth *central moment* M_j are defined as

$$\mu = \frac{1}{n}\sum_{i=1}^{n} x_i, \tag{7.1}$$

$$M_j = \frac{1}{n}\sum_{i=1}^{n} (x_i - \mu)^j, \tag{7.2}$$

where n is the number of points in the time history record. The number of degrees-of-freedom for a given type of variable distribution is defined by the number of moments required for its unique description [16]. For example, for a Gaussian distribution, only two moments, the mean value μ and variance σ^2, suffice for a complete definition of the shape of the PDF; therefore Gaussian distributions exhibit two degrees of freedom.

Two parameters, the *kurtosis* k_u and the *skewness* s_k, are the principal metrics describing the non-Gaussian features of the PDF. They are expressed in terms of the central moments as

$$\begin{aligned} k_u &= \frac{M_4}{M_2^2} = \frac{M_4}{\sigma^4}, \\ s_k &= \frac{M_3}{M_2^{3/2}} = \frac{M_3}{\sigma^3}. \end{aligned} \tag{7.3}$$

The kurtosis characterizes the sharpness of the PDF peak and the width of the PDF tails. The skewness is a measure of the asymmetry of the PDF. The Gaussian process has the kurtosis $k_u = 3$; the process is regarded as *leptokurtic* if $k_u > 3$ and *platykurtic* if $k_u < 3$ [17]. Both kurtosis and skewness of a stress signal affect the damage-accumulation rate and consequently the fatigue life. Indeed, under the hypothesis of a random uniaxial stress state, a process distribution with $k_u > 3$ determines a fatigue damage higher than the Gaussian stress history, because the longer tails cause fatigue cycles with greater amplitude [60, 94]. When relating vibration fatigue to non-Gaussian signals in Chapter 12, the main focus is given on the kurtosis value of the excitation and the stress response, since it was shown in [94] that the influence of the kurtosis is much higher than that of the skewness.

7.2 Identifying nonstationary signal

By identification and quantification of nonstationarity of a signal we can determine if a signal can be processed and analyzed with the classic theory of fatigue analysis as if it was stationary (even if it is not). In the literature [15, 136,137] a number of research methods of evaluating the nonstationarity of a process are explained. Within this book, we evaluate the nonstationarity using the *run-test* [135].

The run test is a nonparametric method based on the idea of dividing the signal to be analyzed into time-windows, and for every window, calculating the variation of one of the statistical properties with respect to the same property of the entire signal [135]. It is based on the definition of a *run* as a sequence of identical observations followed and preceded by a different observation or no observation at all [15]. This means that every window is assigned a value, for example, 1 or 0, depending on the relationship between the statistical characteristics of the windowed signal and the conditions imposed by the test, and then the runs are evaluated. With the run-test approach, too few or too many of the runs can be a proof of nonstationarity [138]. Hence the distribution of the number of the runs in a sequence is a random variable that has a mean μ_r and variance σ_r^2:

$$\mu_r = \frac{2N_1N_2}{N} + 1, \qquad \sigma_r^2 = \frac{(2N_1N_2(2N_1N_2 - N))}{N^2(N-1)}, \tag{7.4}$$

where N_1 and N_2 are the numbers of observations based on the conditions imposed by the test, and N is the total number of observations [135]. Once the level of significance is chosen, the confidence interval is

$$\mu_r \pm \alpha\,\sigma_r, \tag{7.5}$$

where α is the confidence coefficient: considering a level of 95% of confidence, α is equal to 1.96. Thus, if the number of runs falls inside the interval, the signal is supposed to be stationary; otherwise, it should be regarded as nonstationary. At the end, the number of runs can be divided by the expected mean number of runs, defining the *nonstationarity index*

$$\gamma = \frac{r}{\mu_r} \quad [\%]. \tag{7.6}$$

The nonstationarity index indicates the level of nonstationarity of the process: numbers close to 1 should represent a stationary process.

Rouillard [135] defined the run-test

$$V(n) = \begin{cases} 1, & R_{\mathrm{w}}(n) > R_{\mathrm{T}}, \\ 0, & R_{\mathrm{w}}(n) \leq R_{\mathrm{T}}, \end{cases} \tag{7.7}$$

where R_w is the root-mean-square (RMS) value of the windowed signal, R_T is the average mean-square value for the entire sample record, and n is the observation index of window.

Capponi et al. [122] proposed a slightly different criterion which is also used in this book:

$$V(n) = \begin{cases} 1, & |R_w(n) - R_T| > \sigma_R, \\ 0, & |R_w(n) - R_T| \leq \sigma_R, \end{cases} \tag{7.8}$$

where the RMS of the window $R_w(n)$ and the average R_T are limited on both sides by the standard deviation of all the windows

$$\sigma_R = \sqrt{\frac{1}{N_w} \sum_{n=1}^{N_w} (R_w(n) - R_T)^2}, \tag{7.9}$$

where N_w is the total number of windows.

In contrast to the one-sided approach (no absolute value in Eq. (7.8))) the two-sided approach detects higher and also lower values than the average of the entire sample record. The application of both methods on the time signal is presented in Fig. 7.1.

The run-test exhibits a high sensitivity to window width [135]. A short window width can reveal rapid variations of the signal, which may not necessarily represent nonstationarity. On the other hand, if the window width is too wide, then a significant short-duration nonstationarity may not be detected. The influence of window width is numerically and experimentally evaluated in Chapter 12 of Part II.

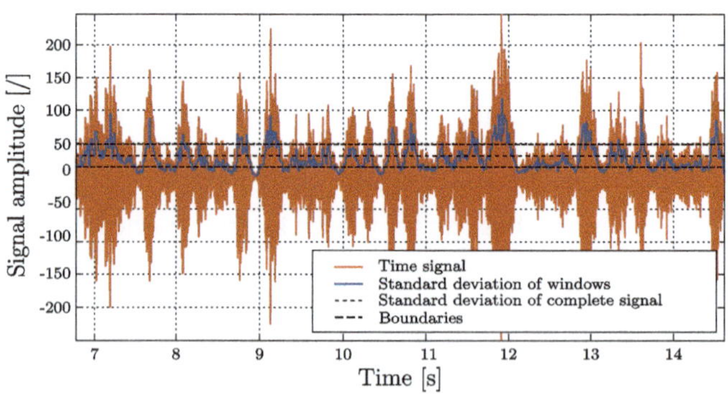

FIGURE 7.1 Run-test evaluation of a time signal.

Part II

Experimental research

Chapter 8

Harmonic accelerated vibration-fatigue testing

In real environment, the products are exposed to different types of vibration loads; in this chapter, we deal with accelerated vibration testing, which is usually prepared with electrodynamic shakers. Instead of a real product, we study a laboratory test sample. The researched *Y-sample* is made from cast aluminium and machined to the specified geometry defined in Fig. 8.1. The Y-sample was designed with regard to the structural dynamics response, which can be significantly altered by attaching additional masses; see Fig. 8.4. Additionally, the fatigue failure location is roughly located at the three locations with smallest cross-section; however, the exact location of failure depends on the excited mode shape (as will be seen later, see, e.g., Ch. 11. To understand the structural dynamics background, we refer the reader to Ch. 1.

The main aim of the test case presented in this chapter is experimental evaluation to what extent do spectral methods (Sec. 4.3) comply with the actual vibration fatigue of real structure when excited with dynamic force (i.e., *dynamic excitation*) or using base motion (i.e., *kinematic excitation*). Besides scientific integrity, a motivation for such test case arises from the vibration tests that are in a product development stage prescribed within environmental test and are defined either in standardized [128,139] or in internal form. It should be noted that the contents of this chapter expands the research [118] and relates it in a more general form to vibration fatigue. As the required theoretical aspects are already given in Sec. 1.2, certain details regarding natural frequency tracking are omitted, and the test results for a larger number of samples are presented later in Ch. 9. For further research, the reader can also check the following references: for dynamic approaches of fatigue testing [118,90,140–145], additional insight into structural dynamics for fatigue testing [12,6,146,147], and damping in particular [148,149].

In the remainder of this chapter, we establish an accelerated vibration fatigue testing methodology, where a test specimen is excited with a harmonic excitation signal at a frequency close to a natural frequency. In Section 8.1 certain preliminary subjects are covered, necessary for later experiment design (Sec. 8.2) and evaluation of obtained experimental results (Sec. 8.3). The

Vibration Fatigue by Spectral Methods. https://doi.org/10.1016/B978-0-12-822190-7.00016-5

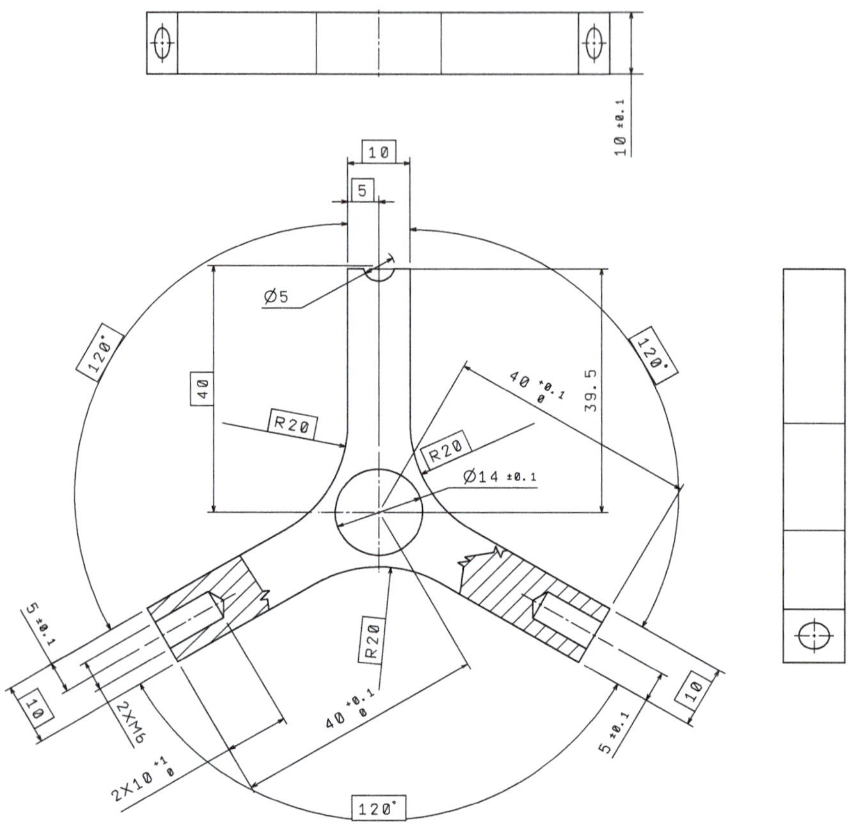

FIGURE 8.1 Drawing of proposed specimen geometry.

presented method is later used in Ch. 9 to obtain the actual material fatigue parameters for aluminium alloy A-S8U3.

8.1 Preliminary analysis of experiment

By exciting dynamic structure close to a natural frequency significant amplifications of strain response are inevitable. In most engineering problems, great efforts are invested to reduce or attenuate any dynamic amplifications of structure. However, in rare cases when it is desirable to achieve high displacements or stress loads of the structure with as little excitation power as possible, a dynamic amplification can be beneficial. The idea of performing accelerated fatigue test on electrodynamic shaker with close-to-resonant excitation coincides with the latter observation. Since the basic principle of such accelerated fatigue testing greatly depends on the specimen dynamic response, certain preliminary checks are required and are described next.

8.1.1 Specimen geometry considerations

For confirming the adequateness of structure's dynamic response, a simple modal analysis of the numerical model of the structure suffices. Besides natural frequencies and corresponding modeshapes, the amplitude of amplification depends also on the damping structure properties, which can be obtained only experimentally. However, for lightly damped structures, such as aluminum Y-sample, low-damping properties do not alter the resonant areas of a structure, therefore diminishing the need for preliminary experimental analysis of Y-sample. Generally, three main guidelines for evaluating the appropriateness of the test specimen geometry should be followed [150]:

1. natural frequency, in vicinity of which a close-to-resonant excitation will be applied, should be well separated from the remaining natural frequencies and should lie within shaker frequency range,
2. corresponding modeshape should be well excited with translational movement in shaker axial direction,
3. fatigue zone with maximal stress loads should be suitable for strain gage installation.

Regardless of a proposed specimen geometry, a numerical modal analysis is a necessary step prior to experiment test and should be performed to confirm the compliance with stated guidelines. Furthermore, numerical analysis reveals the stress state on the fatigue zone, which can be uniaxial (Ch. 4) or multiaxial (Ch. 5). For multiaxial stress state, the numerical model also reveals the ratio of principal stresses σ_1/σ_2, for which the Wöhler curve later obtained is valid.

Fig. 8.2 shows first five modeshapes of Y-sample with corresponding natural frequencies. It should be noted that the weight of a single mass was 52.5 g, which results in relatively high natural frequencies. According to evaluation guidelines, the first and fourth modeshapes are appropriate for accelerated fatigue testing. In the following test cases, only the fourth modeshape was used, since it relates to three times higher natural frequency when compared to the first modeshape, thus resulting in three times shorter duration of the fatigue test. Finally, the principal stress σ_1 distribution in the fatigue zone for the fourth modeshape is shown in Fig. 8.3, and the stress state is practically uniaxial with ratio between the von Mises equivalent stress σ_{eq} and principal stress σ_1 equal to $\sigma_{eq}/\sigma_1 = k_{eq} = 0.971$.

8.1.2 Load stress monitoring during fatigue crack propagation

In classical fatigue tests for obtaining S-N curve of material an amplitude of stress load on structure is controlled and should remain unchanged during a complete test. For duplicating such test conditions on a dynamic structure under close-to-resonant excitation an additional discussion is required. Ideally, with known time-invariant natural frequency and damping ratio of specimen, the stress load does not change during close-to-resonant excitation with constant

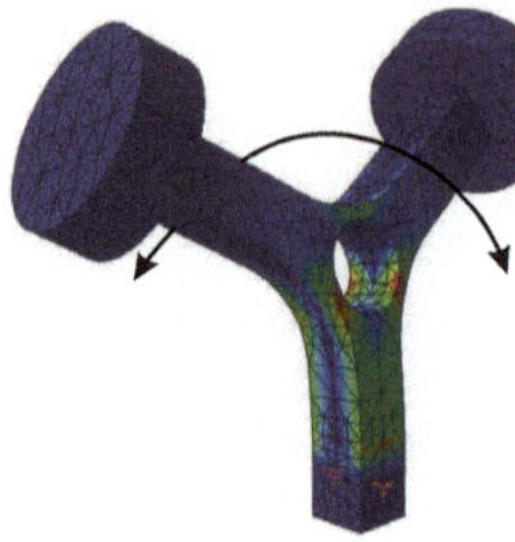

(A) First modeshape, $f_1 = 289.6$ Hz.

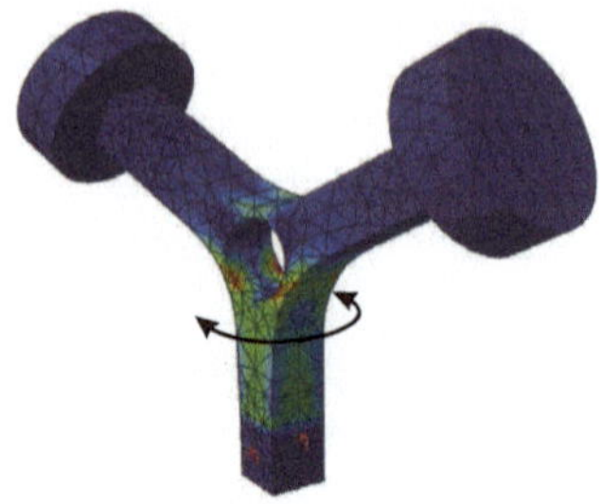

(B) Second modeshape, $f_2 = 344.0$ Hz.

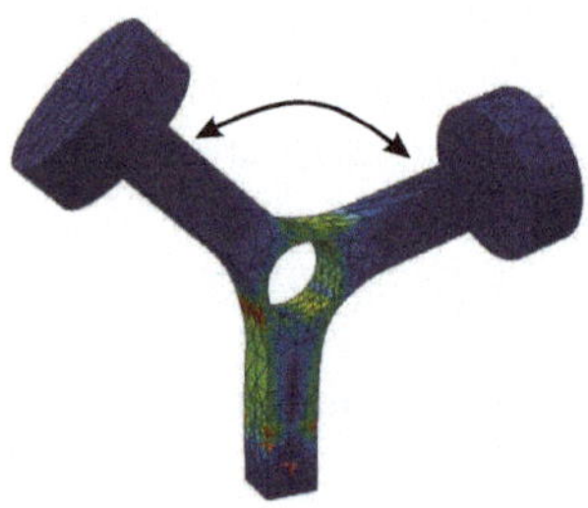

(C) Third modeshape, $f_3 = 425.3$ Hz.

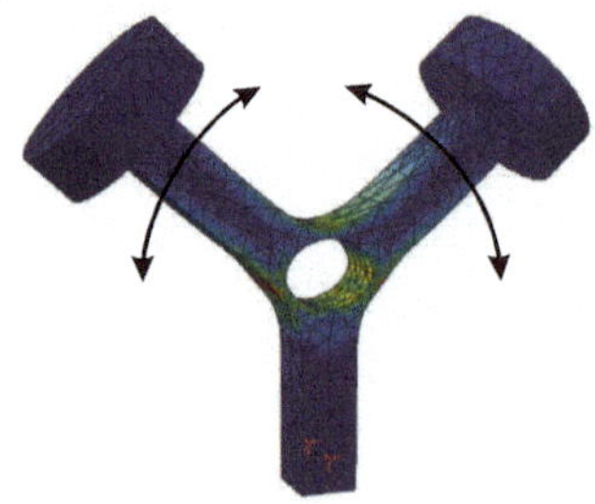

(D) Fourth modeshape, $f_4 = 794.3$ Hz.

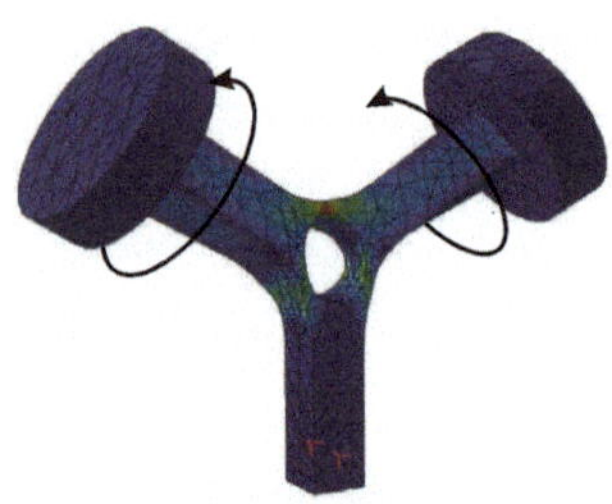

(E) Fifth modeshape, $f_5 = 1594$ Hz.

FIGURE 8.2 First five modeshapes of Y-sample and corresponding distribution of equivalent stress σ_{eq}.

parameters, for example, excitation frequency and amplitude. For such a case, only a simple calibration procedure is required by direct strain gage measurement near the fatigue zone. However, due to occurrence of microcracks during damage accumulation, the structure stiffness and damping properties do change and generally result in a drop of the natural frequency and the damping ratio increases. To overcome this, a real-time strain measurement with strain gage can be performed establishing an active control of stress load on the fatigue zone.

An active strain control may be simplified for linear structure by measuring relative displacement on two locations on the specimen, which shows linear relation to the strain on the fatigue zone (1.205). The latter idea is addressed as indirect strain/stress measurement. In that way the need for strain gage in-

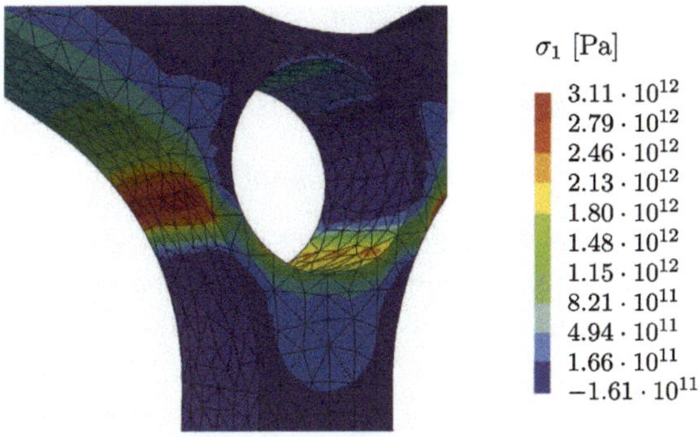

FIGURE 8.3 σ_1 distribution on fatigue zone for the fourth modeshape.

stallation on individual specimen is eliminated, reducing specimen preparation time and testing expenses. In the proposed testing methodology a relative displacement is measured with two single axis accelerometers. The main idea of proposed methodology is based on maintaining a constant relative displacement and phase shift, which ensures method insensibility to variations of natural frequency and damping. For the indirect strain measurement of observed Y-sample, the placements of accelerometers were chosen according to excited modeshape (Fig. 8.2D) and are presented in Fig. 8.4, for example, one on the fixed beam to measure excitation-related acceleration and one on the opposite beam to measure response-related acceleration.

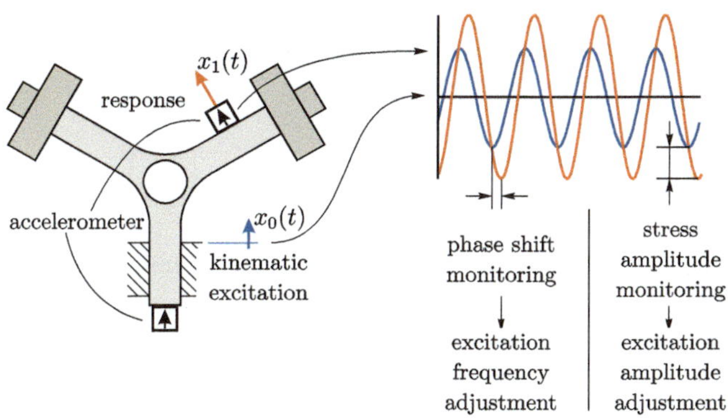

FIGURE 8.4 Experiment setup outline.

Proposed indirect stress measurement relies on fundamental equation of motion for MDOF system (1.182). When a system is excited with kinematic (base) excitation, the excitation force is unknown but has opposite value to the inertial forces of the whole structure. In this case, Eq. (1.182) becomes

$$\mathbf{M}\ddot{\mathbf{x}} + \mathrm{i}\mathbf{D}\mathbf{y} + \mathbf{K}\mathbf{y} = \mathbf{0}, \tag{8.1}$$

where $\mathbf{x}$ and $\mathbf{y}$ represent the vectors of absolute and relative displacements, respectively. By knowing the structure geometry and its discretization a relation between relative and absolute displacements is

$$\mathbf{y}(t) = \mathbf{x}(t) - \mathbf{g}\,x_0(t), \tag{8.2}$$

where $\mathbf{g}$ is the direction influence vector, and $x_0(t)$ denotes the base motion. Under the assumption of harmonic excitation and response displacement (Secs. 1.1.2 and 1.2.3), the transmissibility for such kinematically excited system translates to

$$\mathbf{Q}(\omega) = \frac{\mathbf{y}(\omega)}{x_0(\omega)} = \omega^2 \left(\mathbf{K} + \mathrm{i}\,\mathbf{D} - \omega^2\mathbf{M}\right)^{-1} \mathbf{M}\,\mathbf{g}. \tag{8.3}$$

By introduction of proportional damping model (1.154) and observing only a single ath degree-of-freedom response the transmissibility function can be written as [24]

$$\frac{Y_a}{X_0}(\omega) = \mathbf{m}_a\,\mathbf{g}\,\omega^2 \sum_{r=1}^{N} \frac{\phi_{ar}}{\omega_r^2 - \omega^2 + \mathrm{i}\eta_r\omega_r^2}. \tag{8.4}$$

In Eq. (8.4), Y_a and X_0 denote the amplitudes of response and excitation displacement, respectively, $\mathbf{m}_a$ is the ath row of the mass matrix $\mathbf{M}$, and ϕ_{ar} is the value of the ath element for the rth mass-normalized modeshape (1.124). The damping ratio η_r is defined as in Eq. (1.158), where ϵ and ν present constants in the proportional damping model.

Supposing close-to-resonant excitation of well-spaced pth natural frequency, a response of Y-sample at the ath degree-of-freedom can be obtained using a simple model of SDOF system (1.54):

$$\frac{Y_a}{X_0} = K_a\,\omega^2 \left(\frac{\phi_{ap}}{\omega_p^2 - \omega^2 + \mathrm{i}\,\eta_p\,\omega_p^2} + B_a\right), \tag{8.5}$$

where the constant B_a includes total contribution of the remaining modes $p \neq r$, the value of K_a can be obtained by multiplication of $\mathbf{m}_a$ and $\mathbf{g}$ (8.4). Similarly, the stress response of dynamic structure at the bth degree-of-freedom can be written as

$$\frac{\sigma_{1,b}}{X_0} = K_{\sigma_1,b}\,\omega^2 \left(\frac{\sigma_1\phi_{bp}}{\omega_p^2 - \omega^2 + \mathrm{i}\,\eta_p\,\omega_p^2} + B_{\sigma_1,b}\right). \tag{8.6}$$

Returning to a real structure of Y-sample, by measuring the amplitudes of absolute accelerations and consequently absolute displacements X_0 in X_1 from Eq. (1.21) (Fig. 8.4) the amplitudes of relative displacement between measurement positions can be obtained by

$$Y_1 = |X_1 - X_0|\,. \tag{8.7}$$

Obviously, the relative displacement may be regarded as a known quantity during fatigue test. Considering Eqs. (8.5) and (8.6), we easily obtain the ratio between the fatigue zone stress and relative displacement:

$$\frac{\sigma_1}{Y_1}(\eta_p, \omega_p) = \frac{\left| \frac{\sigma_1 \phi'_p}{\omega_p^2 - \omega^2 + \mathrm{i}\, \eta_p\, \omega_p^2} + B'_{\sigma_1} \right|}{\left| \frac{\phi'_{1p}}{\omega_p^2 - \omega^2 + \mathrm{i}\, \eta_p\, \omega_p^2} + B'_1 \right|}. \tag{8.8}$$

This equation omits the labeling of degrees-of-freedom, since the position of response measurement and fatigue zone on the Y-sample are known. In Eq. (8.8), $'$ denotes appropriate multiplication by the mass matrix $\mathbf{M}$ and constants related to the direction vector $\mathbf{g}$. Obviously, the only remaining variables are ω_p and η_p, which may alter the ratio σ_1/Y_1 used for indirect stress measurement.

Detailed analysis on what influences do changes of ω_p and η_p have on the ration σ_1/Y_1 was performed in [118] with validated numerical model. Based on the results, shown in Fig. 8.5, the proposed indirect stress measurement does not show relevant sensitivity to changes of natural frequency ω_4 and damping ratio η_4. Accordingly, for the case close-to-resonant excitation, throughout the whole accelerated fatigue test, the following conclusion is justified:

$$\frac{\sigma_1}{Y_1} \approx \text{const} = k_\sigma. \tag{8.9}$$

This also confirms the validity of indirect stress measurement approach by measuring absolute accelerations on two measurement positions. However, the value of ratio σ_1/Y_1 should be determined prior to the conduction of accelerated fatigue test. Experimental validation of the presented theoretical conclusions is presented in Sec. 8.2.4.

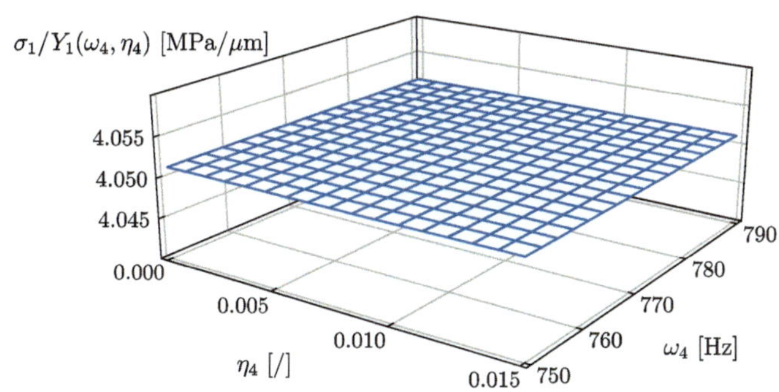

FIGURE 8.5 Ratio $\sigma_1/Y_1(\omega_4, \eta_4)$ during changes of ω_4 and η_4.

8.2 Experiment setup

This section focuses on controlling the dynamic response of Y-sample and shows the results of preliminary measurements, necessary for appropriate conduction of accelerated fatigue test.

8.2.1 Dynamic response control

The accelerated fatigue test was performed on electrodynamic shaker LDS V555, to which the Y-sample was attached with fixture shown in Fig. 11.2. The Y-sample was excited at the base (Fig. 8.4) with harmonic base excitation (Sec. 1.1.5):

$$x_0(t) = X_0\,\mathrm{e}^{\mathrm{i}\,\omega\,t}. \tag{8.10}$$

With such an excitation signal, the dynamic response of Y-sample can be controlled by adjusting excitation amplitude X_0 and excitation frequency ω to maintain the constant stress amplitude at the fatigue zone. However, determining appropriate X_0 and ω is not trivial; changes of ω_4 and η_4 alter dynamic response of Y-sample and consequently require appropriate adjustment of excitation signal $x_0(t)$. To achieve this, a feedback control loop for real-time regulation of ω and X_0 was designed and is presented next.

8.2.2 Excitation frequency control

Assuming a close-to-resonant excitation with excitation frequency ω slightly above the natural frequency ω_4, even a minor reduction of ω_4 and therefore an increase of $|\omega-\omega_4|$ results in lower dynamic amplification. This leads to a lower relative displacement Y_1 and, according to Eq. (8.9), to a lower stress load on the fatigue zone. Logically, for obtaining S-N curve (Eq. (4.1)), this is not acceptable. Additionally, by increasing $|\omega-\omega_4|$ the influence of the remaining modes can arise. To prevent such cases, an excitation frequency ω should be controlled in real-time throughout the whole fatigue test.

For harmonic excitation (Eq. (8.10)) of linear SDOF system, a response signal can be written as in Eq. (1.21):

$$x_1(t) = X_1\,\mathrm{e}^{\mathrm{i}\omega\,t}. \tag{8.11}$$

Active excitation frequency ω control can be performed by maintaining constant phase shift φ between excitation $x_0(t)$ and response $x_1(t)$ signal [135]. Fig. 8.6 depicts a constant phase shift approach for the case of two transmissibility functions of Y-sample, one being measured on undamaged specimen and the other on a specimen after being exposed to a certain number of load cycles. Evidently, once the natural frequency ω_4 drops, the excitation frequency ω should be reduced accordingly to maintain the constant phase shift φ.

Due to harmonic nature of $x_0(t)$ and $x_1(t)$, the actual phase shift φ can be determined on time domain signals as shown in Fig. 8.4. Ideally, by maintaining the phase shift $\varphi = 90°$ the Y-sample would be excited in exact resonance increasing the exploitation of dynamic amplification. This approach is possible for structures with low damping [135]; for the case of Y-sample with low damping, the excitation in exact resonance would result in unstable response, which would prevent the control of stress amplitude σ_1. Therefore a phase shift of $\varphi = 175°$ was adopted for accelerated fatigue test of Y-samples.

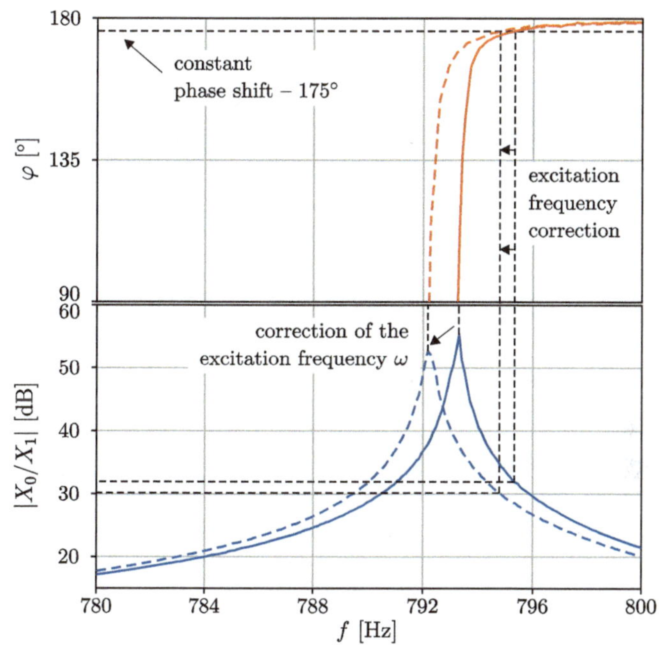

FIGURE 8.6 Y-sample transmissibility at initial undamaged state (——) and after certain number of load cycles (— — —).

8.2.3 Excitation amplitude control

Although the phase shift φ is maintained constant during fatigue test, this does not ensure a constant stress amplitude σ_1 during fatigue test. The occurrence of fatigue microcrack inherently leads to increase of the structure damping ratio η_4, reducing the response-to-excitation ratio Y_1/X_0 and consequently reducing the stress amplitude σ_1. The changes of relative displacement Y_1 for constant excitation amplitude X_0 due to changes of natural frequency ω_4 and damping ratio η_4 were calculated for SDOF model of Y-sample in [118] and are shown in Fig. 8.7.

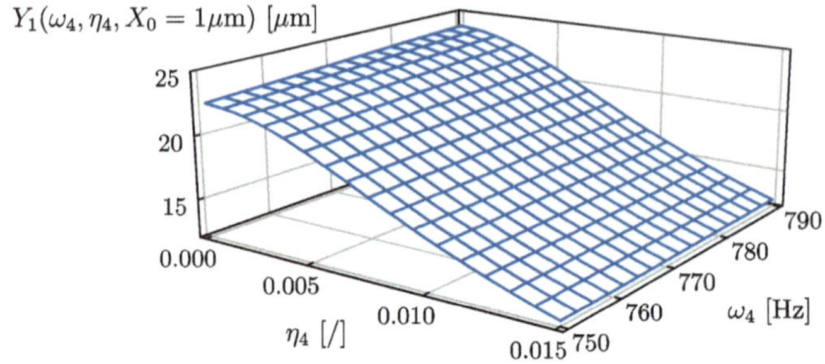

FIGURE 8.7 Changes of relative displacement during changes of natural frequency ω_4 and damping ratio η_4.

Obviously, besides the phase angle φ, the excitation amplitude X_0 should also be controlled in real time during the accelerated fatigue test to maintain the constant stress amplitude σ_1. The exact value of required excitation amplitude X_0 is defined by indirect measurement of the stress amplitude.

8.2.4 Verification of indirect stress measurement

Before any of the accelerated fatigue tests were conducted, an experimental verification of the indirect stress measurement was performed. The experimental verification was performed in two steps: firstly, the ratio σ_1/Y_1 was measured for the three Y-shaped specimens (S1, S2, and S3) before the accelerated fatigue test, and secondly, the ratio σ_1/Y_1 was monitored during the accelerated fatigue test of the Y-sample S4 to confirm the constant value σ_1/Y_1 when the modal parameters ω_4 and η_4 begin to change.

Prior to the calibration process, the nonuniform stress state in the fatigue zone was analyzed. The stress state in the fatigue zone can be considered as uniform only over the small area at the location of the maximum principal stress σ_1; see Fig. 8.3. This implies that a very small strain gage should be used. However, only a slight mislocation of the small strain gage could cause moderate measurement errors and issues with repeatability; therefore a larger strain gage was used for calibration. Because of the nonuniform stress state on the surface, covered by a larger strain gage, it is necessary to obtain the ratio between the maximum principal stress σ_1 and the average principal stress $\sigma_{1\ \text{average}}$ (the actual value being measured). For the calibration experiment, the ratio $\sigma_1/\sigma_{1\ \text{average}}$ was obtained for the strain gage with a 6-mm measuring grid; see Fig. 8.8. This was done numerically with modal analysis of the finite-element model and experimentally with two strain gages with measuring grids of 6 mm and 0.6 mm, respectively. The stress measured with the 0.6-mm strain gage was considered as σ_1. The agreement between the numerical and experimental results was within 1.5% and the ratio $\sigma_1/\sigma_{1\ \text{average}} = 1.31$ was determined.

Fig. 8.9 shows the relation between the measured principal stress σ_1 and the relative displacement Y_1 for the three selected Y-shaped specimens before the accelerated fatigue test was conducted, when no changes of ω_4 and η_4 were present. Fig. 8.9 confirms the linear relation between the principal stress σ_1 and the relative displacement Y_1 for the constant natural frequency ω_4 and the damping loss factor η_4. The agreement between the theoretically (Fig. 8.5) and experimentally (Fig. 8.9) obtained values for the ratio σ_1/Y_1 is very good. The trend-line slopes for the linear regression between the specimens S1–S3 differ by less than 2%, and therefore a good repeatability of the indirect stress measurement is confirmed.

FIGURE 8.8 Calibration experiment setup.

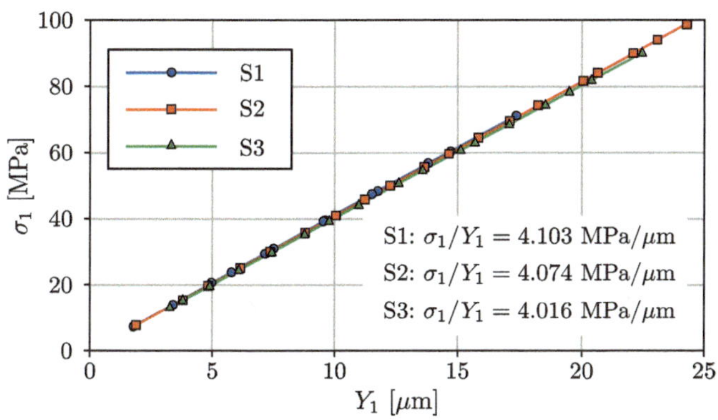

FIGURE 8.9 Calibration of specimens S1–S3.

Fig. 8.10 shows the ratio σ_1/Y_1 during the accelerated fatigue test of the specimen S4. In this particular experiment the principal stress σ_1 and the relative displacement Y_1 were monitored until the failure of the strain gage occurred after $8 \cdot 10^5$ load cycles (approximately 70% of the specimen life-time). As predicted by the theoretical analysis in Sec. 8.1, we can observe in Fig. 8.10 that during the experiment the measured ratio σ_1/Y_1 remained constant until the strain gage failure, although the changes in the natural frequency ω_4 and η_4 were present. According to the presented measurements in Figs. 8.9 and 8.10, the indirect stress measurement offers an accurate monitoring of the principal stress σ_1 during the whole accelerated fatigue test and is not exposed to failure, as in the case of the strain gage. Additionally, the indirect stress measurement eliminates the measurement error due to the inaccurate strain gage positioning and greatly simplifies the experimental setup and the specimen preparation.

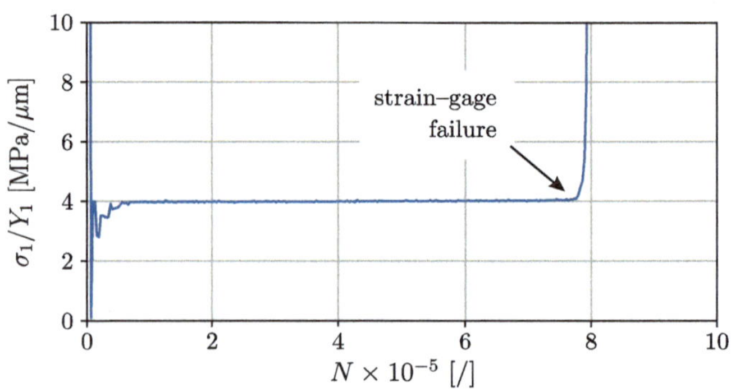

FIGURE 8.10 The ratio σ_1/Y_1 during the accelerated fatigue test of the specimen S4.

8.3 Accelerated fatigue test

The pilot accelerated fatigue test was performed on the Y-sample above the resonant area at the constant phase angle $\varphi = 175°$ and at the principal stress amplitude in the fatigue zone of $\sigma_1 = 125$ MPa. The aim of this fatigue test is verifying the realization of the uninterrupted accelerated fatigue test by monitoring the actual principal stress σ_1 and the phase angle ϕ simultaneously with the real-time control (close to the resonance).

The measurements, shown in Figs. 8.11, 8.12, and 8.13, confirm the adequacy of the controller software, since the controlled quantities φ and σ_1 remained constant, although the natural frequency of the specimen reduced by 10 Hz and the damping loss factor increased by almost 100% during the whole fatigue process. Even when the specimen was harmonically excited only 3 Hz from the natural frequency, the whole experiment was stable and well controlled. Additionally, the measurement of the temperature at specimen fatigue zone showed that the temperature changes due to the structural damping were negligible compared to the changes due to the heat conduction from the shaker (approximately 5°C).

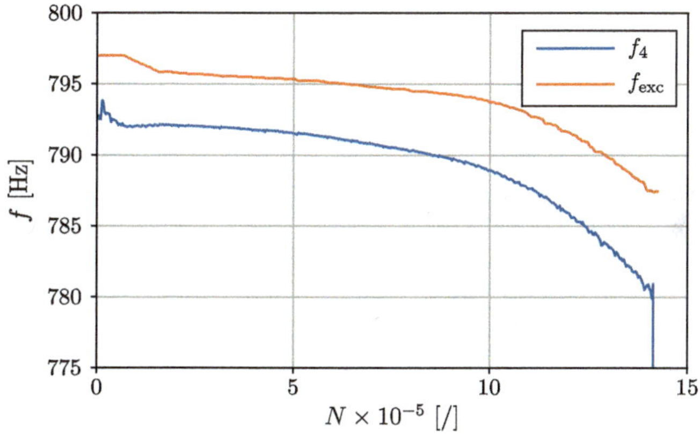

FIGURE 8.11 Excitation frequency f and identified natural frequency f_4 of the tested specimen during the pilot run of the accelerated fatigue test.

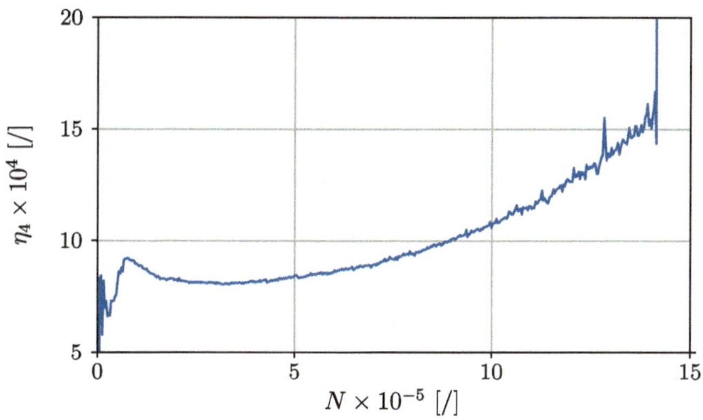

FIGURE 8.12 Damping loss factor η_4 during the pilot run of the accelerated fatigue test.

The final fatigue failure occurred after $1.4 \cdot 10^6$ load cycles, which took only 30 min to accomplish. The fatigue crack is shown in Fig. 8.14 and occurred exactly in the critical fatigue zone, as predicted by the numerical modal analysis. Contrary to the steady natural-frequency drop observed in Sec. 9.2 for random excitation, a sudden drop of natural frequency and change of current phase shift φ occurred for the case of harmonic excitation. For accelerated vibration fatigue

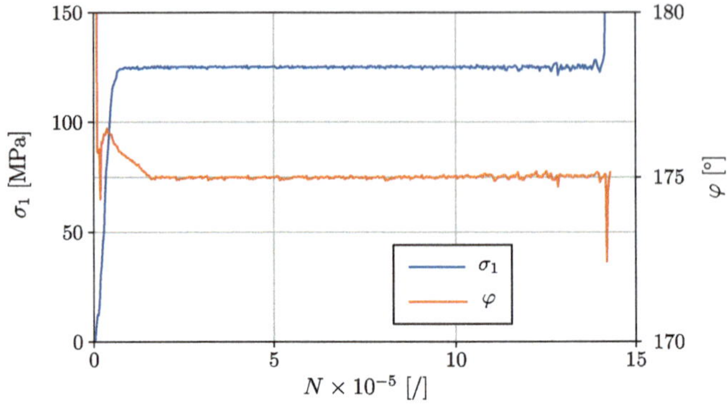

FIGURE 8.13 Principal stress amplitude σ_1 and phase angle φ during the pilot run of accelerated fatigue test.

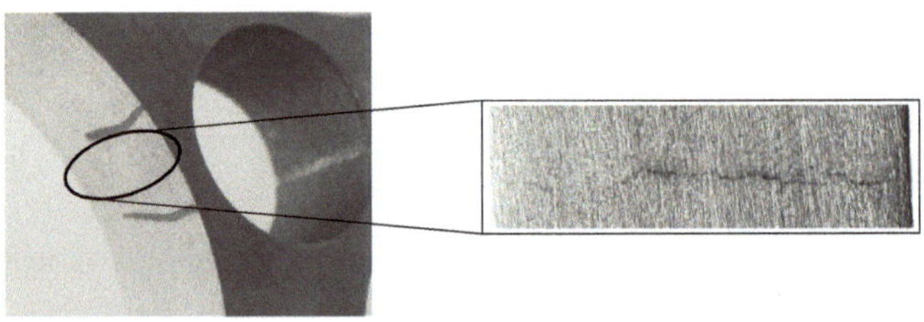

FIGURE 8.14 Fatigue crack on the tested specimen.

test, the specimen time-to-failure was therefore declared at the occurrence of instantaneous change of the phase angle.

Chapter 9

Harmonic vs. random excitation

In Ch. 8, we discussed the methodology for harmonic accelerated fatigue-testing. In this chapter, we discuss the harmonic in comparison to the narrow-band excitation and describe the random excitation level with PSD level of the narrow-band profile, whereas in Ch. 10, we will generalize the random uniaxial loads to the frequency-domain fatigue-damage counting methods.

We will conduct the experiments on the Y-sample presented in Ch. 8 (material A-S8U3).

9.1 Fatigue tests with harmonic excitation

For conducting accelerated fatigue tests on Y-samples, nine specimens were prepared, H1–H9. During single test-run, all parameters, shown in Figs. 8.11–8.13, were measured for particular specimen. Due to excitation stability, Y-samples were excited below resonance by maintaining the reference phase shift $\varphi_{\mathrm{ref}} = 3.5^\circ$.

Samples H1–H9 were manufactured in another lot compared to specimens S1–S3, which resulted in different value of the fourth natural frequency ω_4 and the corresponding damping ratio η_4. The observed differences are a result of variations in the chemical compound of a material, temperature conditions during sand casting, and surface finish with wire EDM. Consequently, a calibration measurement (Sec. 8.3) was repeated, and a similar value of the ratio $Y_1/\sigma_1 = 3.9$ MPa/µm was identified.

As expected, during damage accumulation, a significant drop of natural frequency ω_4 occurred together with an increase of damping ratio η_4, as presented in Figs. 9.1 and 9.2. All experimental results are summarized in Table 9.1, where $f_{4\,\mathrm{init}}$ denotes the initial fourth natural frequency of undamaged specimen, $\min(\eta_4)$ is the lowest value of the identified damping ratio, and Δf_4 is the total change of the fourth natural frequency before final failure of the specimen.

According to Fig. 9.1, the changes of the fourth natural frequency do not result in the similar values, even when the excitation level remains the same. This justifies a declaration of final failure by observing sudden change of natural frequency. Additionally, Fig. 9.2 shows significant differences in damping ratios between tested specimens; higher loads resulted in higher damping. This was attributed to the additional damping source, which arises locally at the contact between Y-sample and fixture.

The experimental times-to-failure with respect to the applied stress loads are shown on the σ–N diagram, see Fig. 9.3. The scatter of experimental results is

Vibration Fatigue by Spectral Methods. https://doi.org/10.1016/B978-0-12-822190-7.00017-7

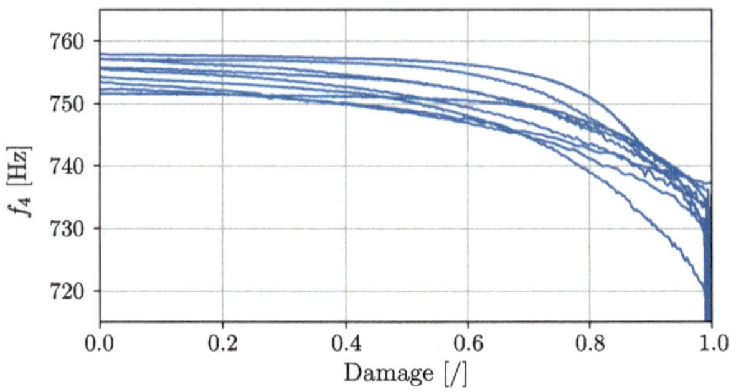

FIGURE 9.1 Natural-frequency drop of specimens H1–H9.

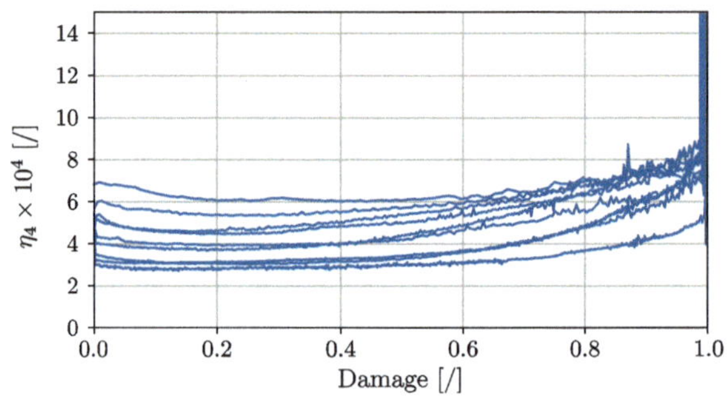

FIGURE 9.2 Damping ratio increase of specimens H1–H9.

TABLE 9.1 Experimental results for obtaining A-S8U3 fatigue parameters.

Specimen	σ_1 [MPa]	Cycles-to-failure N	$f_{4\,\mathrm{init}}$ [Hz]	$\min(\eta_4)\times10^4$	Δf_4 [Hz]
H1	95	$33.0{\cdot}10^5$	758.0	3.10	31.3
H2	105	$25.0{\cdot}10^5$	751.6	2.85	18.3
H3	105	$13.5{\cdot}10^5$	757.2	3.08	28.6
H4	115	$9.71{\cdot}10^5$	757.3	4.49	21.2
H5	115	$13.1{\cdot}10^5$	754.3	3.73	34.8
H6	125	$6.87{\cdot}10^5$	755.8	3.96	21.9
H7	125	$7.98{\cdot}10^5$	755.7	4.60	22.2
H8	135	$4.87{\cdot}10^5$	752.3	5.37	20.9
H9	135	$3.24{\cdot}10^5$	753.4	6.09	16.2

comparable to scatter at standard fatigue tests [37], confirming the adequacy of proposed accelerated-testing methodology. Material fatigue parameters k and C (4.1) were identified with least-square method for linear regression on logarithmic scale [37]. The obtained Wöhler curve for material A-S8U3 is described by (4.1),

$$1.45 \cdot 10^{18} = \sigma^{5.88} N, \tag{9.1}$$

and is also presented in Fig. 9.3.

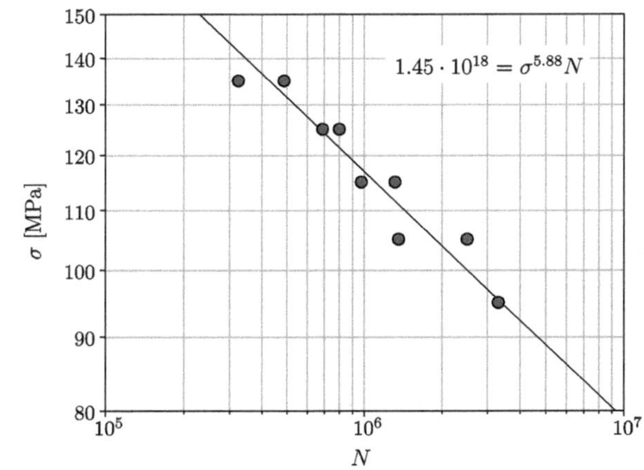

FIGURE 9.3 S-N diagram and corresponding Wöhler curve.

The published values of fatigue parameters for A-S8U3 vary reasonably between the authors [151–155]. Wöhler curve, as shown in Fig. 9.3, is in good correlation with the results obtained by Kurth et al. [151] and shows higher fatigue resistance when compared to results of the González et al. [155] and Dietrich and Radziejewska [154]. Considering the results of ASM International [152], which present an international reference, the obtained Wöhler curve (9.1) shows a lower fatigue resistance. However, this observation is not unexpected, since it is known that the wire EDM surface finish negatively influences the fatigue resistance of material [156].

9.2 Fatigue tests with random-signal excitation

In this section, we present a second experiment, where Y-samples are excited with the random acceleration profile, see Sec. 2.2.8. Here a constant random-signal acceleration profile is maintained at the fixation of the specimen during complete vibration test until the fatigue failure occurs. In this way the conditions during a standard vibration test [128,139] are reproduced.

9.2.1 Experiment setup and results

Experiment setup of the fatigue test with the random base excitation is identical to the setup, illustrated in Fig. 8.8, except that no strain gages were used to monitor stress levels in the fatigue zone. The test was performed on the LSD V875 shaker, and the acceleration profiles had constant level of the acceleration PSD (Sec. 2.2.4) in the frequency band from 600 to 800 Hz. The selection of a narrow band random profile in the vicinity of a single natural frequency greatly reduces the influence that the remaining modes have on the system dynamic response and damage accumulation. However, the frequency band of excitation should be wide enough to ensure the appropriate excitation of the observed mode-shape even when the drop of natural frequency occurs. The excitation of the specimen was controlled with a Dactron Laser controller. Ten Y-shaped specimens R01–R10 were tested: at the start of each test the initial transmissibility was measured with very low acceleration profile to obtain initial damping, and afterward the fatigue test was performed. During the fatigue test, only the transmissibility magnitude between response and control accelerometer (Fig. 8.8)) was measured, from which the changes of natural frequency were identified. The results of the experiment are summarized in Table 9.2.

TABLE 9.2 Experimental data of random-excitation fatigue test.

Specimen	PSD level $[(m/s^2)^2/Hz]$	Time-to-failure [s]	$\eta_{4\,\mathrm{init}} \times 10^4$	Δf_4 [Hz]
R01	0.725	$391 \cdot 10^2$	4.07	71.7
R02	0.725	$316 \cdot 10^2$	5.07	55.6
R03	1.08	$65.0 \cdot 10^2$	4.31	140.4
R04	1.08	$98.7 \cdot 10^2$	4.64	43.7
R05	1.60	$27.4 \cdot 10^2$	3.80	139.7
R06	1.60	$18.8 \cdot 10^2$	7.10	60.8
R07	2.38	$13.6 \cdot 10^2$	5.39	75.6
R08	2.38	$15.2 \cdot 10^2$	4.59	114.3
R09	5.25	$3.98 \cdot 10^2$	4.87	75.9
R10	5.25	$3.84 \cdot 10^2$	5.77	62.6

Natural frequency shifts during the random-excitation fatigue tests are shown in Fig. 9.4. From it we can observe that significantly larger natural frequency shifts occur during the random excitation when compared to the harmonic test; see Fig. 9.1. During the harmonic test, a drop of natural frequency occurred just before the crack propagated through the whole cross-section. For that case, the real-time control was not sufficiently reliable, and the fatigue failure was declared at this point. On the other hand, when the excitation is a random process, the fatigue crack propagates continuously until the full cross-section crack is reached. Since many authors [73,157] assume that frequency shifts relate to the accumulated fatigue damage, the fatigue failure of the specimen under random excitation was declared when a certain value of the natural frequency shift was observed. The critical natural frequency shift for the final fatigue failure at the random-excitation test was 23.9 Hz or, in terms of Eq. (11.1),

$Z = 6.3\%$, which coincides with the average value of the total frequency shift, measured for the specimens at the harmonic tests (see Table 9.1). Accordingly, the times-to-failure of tested specimens are for different PSD levels shown in Fig. 9.5.

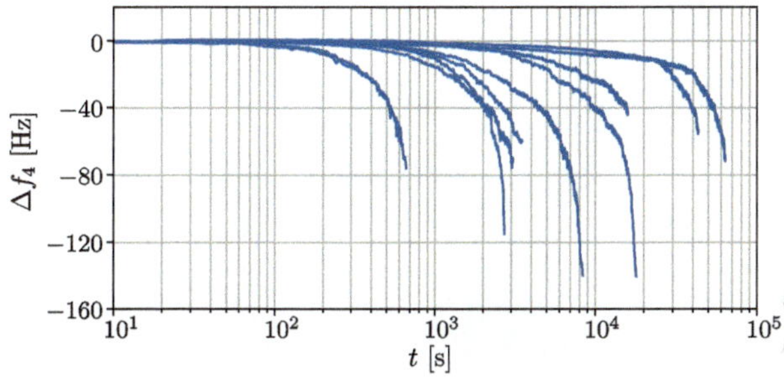

FIGURE 9.4 Fourth natural frequency drop Δf_4 during the random-excitation test.

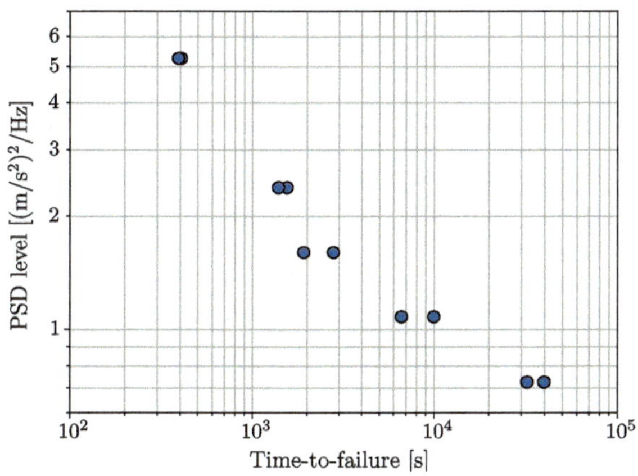

FIGURE 9.5 Times-to-failure for specimens excited with random signal.

9.2.2 Numerical model

The vibration fatigue life of the single tested specimen was obtained in the custom developed Vibration-fatigue plug-in for the Catia analysis environment. The FEM model consisted of 15600 10-node tetrahedral solid elements. The size of the elements in the fatigue zone was approximately 1.5 mm, which provided reliable results of stress responses used for the fatigue life estimation. For the fatigue life estimation, the material parameters from harmonic test (Sec. 9.1) along with the calculated nodal values of equivalent stress from numerical modal analysis were used. Additionally, the initial damping loss factors (Table 9.2) were used to predict the $S_c(\omega)$ defined in Eq. (5.29). In Fig. 9.6 the fatigue life estimation for specimen R02 is visualized. To reduce the calculation time, the fatigue life calculation was performed for the nodes in the fatigue zone only.

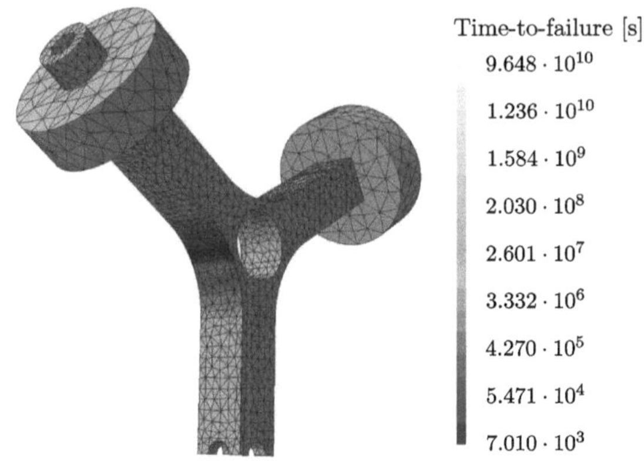

FIGURE 9.6 Predicted time-to-failure for specimen R02.

9.3 Comparison of the experimental and numerical results

The comparison between the estimated and actual times-to-failure for the case of random excitation is shown in Fig. 9.7. From this figure we can notice that the numerical analysis estimated reasonably shorter fatigue lives compared to the actual fatigue lives. This deviation originates from the damping loss factor, proposed in the numerical calculation, which was measured for the intact specimen at a low excitation level. From the discussion regarding the results of the harmonic test the damping loss factor increases with the accumulated fatigue damage and with the higher excitation level (due to the type of specimen fixation). In any case the fatigue life estimation gives conservative prediction when the initial damping loss factor is used.

Additionally, for the tested series of specimens, the effective damping loss factor was obtained with numerical minimization that gives a minimal sum of deviations between actual T_{act} and estimated T_{est} times-to-failure of N tested samples:

$$\Delta T(\eta_4) = \sum_{i=1}^{N} \left(\log_{10}(T_{\mathrm{act},i}) - \log_{10}(T_{\mathrm{est},i}(\eta_4))\right)^2. \tag{9.2}$$

The numerical optimization was performed by the simplex method. The value of effective damping loss factor was identified as $\eta_{4\,\mathrm{eff}} = 9.594 \cdot 10^{-4}$. The comparison between experimental and estimated times-to-failure using the updated value of damping loss factor is shown in Fig. 9.8.

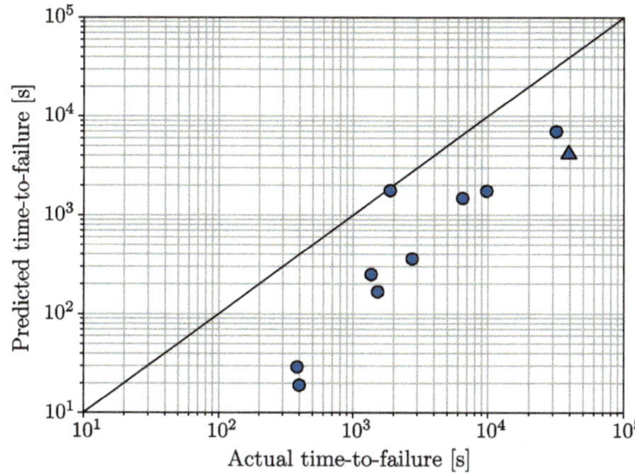

FIGURE 9.7 Comparison of random excitation experiment with numerical prediction using the initial damping loss factors $\eta_{4\,\text{init}}$.

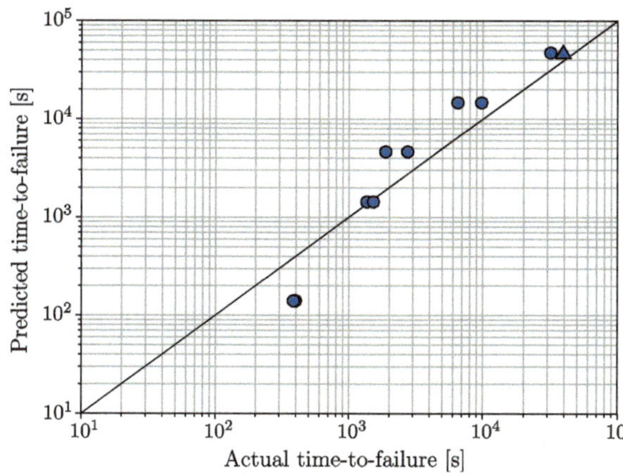

FIGURE 9.8 Comparison of random excitation experiment with numerical prediction using the effective damping loss factors $\eta_{4\,\text{eff}}$.

From the presented comparison we can conclude that the structure damping properties greatly influence the numerical estimation of time-to-failure. Observing the specimen R01 (marked with ▲ in Figs. 9.7 and 9.8): if the value of damping loss factor η_4 is increased by factor of 2.35, then the predicted time-to-failure increases by a factor of 11.3. The presented numerical and experimental results clearly show that uncertainties of the time-to-failure estimation arise from the inherent changes of system modal properties during the damage accumulation and consequential crack propagation. However, by adopting the initial damping loss factor, obtained by experimental modal analysis at low excitation levels, the fatigue life estimation is always on the conservative side.

Chapter 10

Frequency domain methods for fatigue life estimation

This chapter and the subsequent experimental comparison are based largely on the scientific paper by Mršnik et al. [66]. The numerical analysis was updated and redone. The changes are with regards to those spectral methods that base on a simple correction of the narrowband formulations. Whereas in the original paper the expected peak occurrence ν_p (Eq. (2.66)) was used, valid for a broadband process, here we used the expected zero-crossing frequency ν_0 (Eq. (2.62)). The latter is commonly used for narrowband processes but gives more accurate results when used with the narrowband correction methods, compared to the ν_p. Specifically, results were notably improved for the narrowband (Sec. 4.3.1), Wirsching–Light (Sec. 4.3.2), and $\alpha_{0.75}$ methods (Sec. 4.3.3). Furthermore, we performed the revised analysis using higher sampling frequency.

The markedly empirical nature of the models that are spectral moments methods means that an uncertainty exists in how accurately an arbitrary spectrum, not present in the simulations training set, can be evaluated. Therefore, to gain confidence in these methods, a broad set of different spectra typically occurring in structural dynamic analysis were evaluated using spectral methods in parallel with the reference time-domain rainflow counting (Sec. 4.2.2).

Some other comparison studies also focused on the spectral moments and can be found in [79,72,158–163], but are mostly constrained to a smaller set of methods and numerical simulations only.

10.1 Experiment

Producing a group of proper response spectra (5.12) requires actual structures varying in geometry and material. For this comparison, we simulated the spectra directly on the electrodynamic-shaker fixation head, as shown in Fig. 10.1, which meant good control over the generated spectra in a viable experiment setup. A time history was acquired using an accelerometer, and the rainflow counted cycle distribution (Sec. 4.2.2) was compared with spectral estimates (Sec. 4.3) based on the vibration profile, specified in the shaker control software. Comparison of the two estimates from the time and frequency domain gave insight into how different spectral moments method perform in terms of accuracy and precision.

Vibration Fatigue by Spectral Methods. https://doi.org/10.1016/B978-0-12-822190-7.00018-9

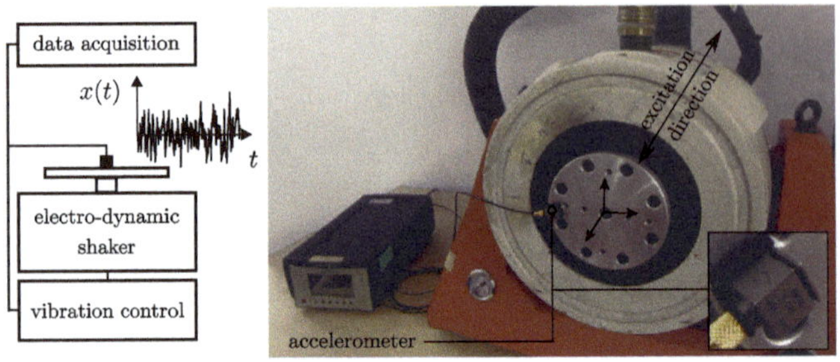

FIGURE 10.1 Experiment setup.

A 5-min sample of time history was recorded on the accelerometer for each simulated spectra at a 10-kHz sampling rate and 24-bit resolution. Commercial hardware and software packages were used to configure and control the electro-dynamic shaker in a feedback loop. The broader schematic of the experiment is shown in Fig. 10.1. Thus recorded data were assumed to be uniaxial stress signal and were saved for later analysis in the time and frequency domain.

10.2 Simulated spectra

Five groups of spectra were formed, each defined by a specific shape of spectrum, with the goal to thoroughly test the accuracy of spectral moments methods. The groups were representing the following categories: multiple vibration modes (MM), background noise level (BN), spectral width (SW), close-modes spectra (CM), and typical automotive profiles (AM).

The *quality factor* Q (not to be confused with the Dirlik method coefficient) is commonly used to characterize resonance peaks. It conveys information about amplitude and sharpness of the peak and can be calculated from the damping factor ζ: $Q = 1/(2\,\zeta)$. It can also be determined based on the half-power points, f_1 and f_2, which are the frequencies where the peak amplitude drops for one half. If the peak is at the central frequency f_0 and $\Delta f = f_2 - f_1$, then the factor Q can be determined as

$$Q = \frac{f_0}{\Delta f}. \tag{10.1}$$

It will be used consistently in the following to quantitatively describe specific shapes of spectra.

The multimodal group of spectra shown in Fig. 10.2 aims to evaluate the sensitivity of spectral methods to the presence of different number of resonance peaks. In the first spectrum, only one peak is present, and another one is added for each spectrum, up to four. The central frequencies of the peaks are, in this order, 200, 400, 600, and 750 Hz. The peak amplitude is the same for all the peaks, 650 MPa^2/Hz.

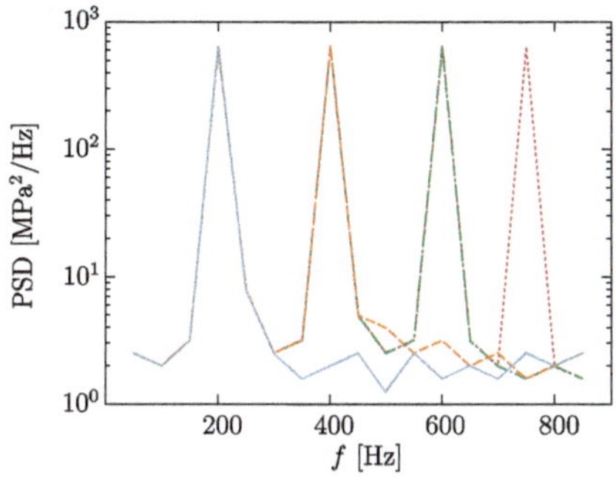

FIGURE 10.2 Multimodal group of spectra.

To test the sensitivity to different background noise level, a bimodal spectrum was devised, characterized by peaks of central frequencies $f_1 = 150$ Hz and $f_2 = 600$ Hz with $Q_1 = 5.00$ and $Q_2 = 17.96$. The background noise level is increased in intervals of 20 MPa/Hz of amplitude. The last increase is when the noise covers the lower peak completely, amounting to seven spectra in total. The group is shown in Fig. 10.3.

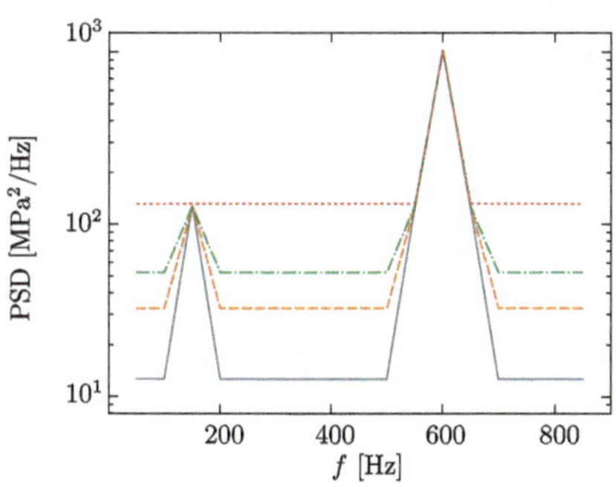

FIGURE 10.3 Background noise level spectra.

The third group compares peaks of different spectral width in terms of accuracy of the spectral moments methods. The sharp peak from the first spectrum in the group, with the central frequency $f_1 = 430$ Hz and $Q = 8.96$, is expanded by 50 Hz left and right, with every additional spectrum. The largest and final bandwidth in the progression, shown in Fig. 10.4, is 450 Hz. This group counts nine spectra in total.

Impact of the closeness of two modal peaks on the time-to-failure estimates is judged based on a collection of close-modes spectra, shown in Fig. 10.5. Starting with two peaks of central frequencies $f_1 = 150$ Hz and $f_2 = 700$ Hz and quality factors $Q = 6.00$ and $Q = 24.00$, the distance between them is reduced by 100 Hz for each additional spectrum, yielding four spectra in total.

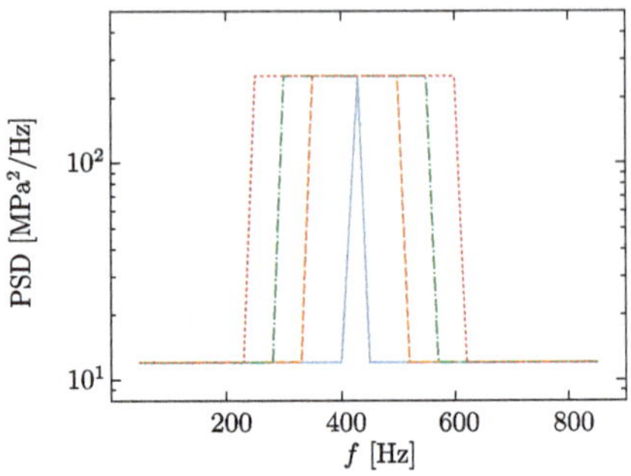

FIGURE 10.4 The spectral-width group of spectra.

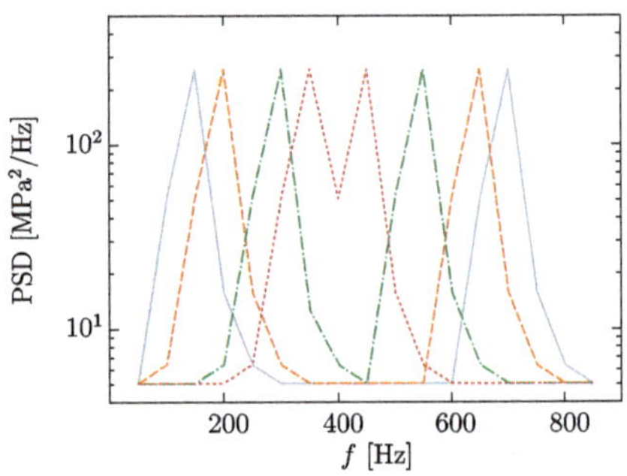

FIGURE 10.5 Close-modes spectra.

Lastly, some attention was given to the accelerated vibration test profiles, commonly used in automotive industry. Three spectra, shown in Fig. 10.6, were simulated on the shaker and participated in the experimental side-by-side comparison of spectral moments methods.

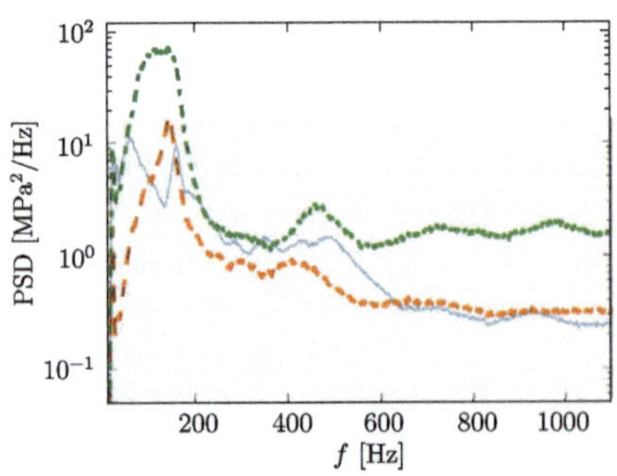

FIGURE 10.6 Typical accelerated vibration test profiles.

10.3 Results

The spectral method estimates based on the vibration profiles (spectra) were compared to the reference approach T_{RFC} obtained on measured data in the time domain, using the rainflow cycle counting method (Sec. 4.2.2) and the Palmgren–Miner damage accumulation hypothesis (Eq. (4.7)):

$$T_{\text{err}} = \frac{T^{\text{XX}} - T_{\text{RFC}}}{T_{\text{RFC}}}. \tag{10.2}$$

The estimate T^{XX} is pertaining to a chosen spectral moments method (Sec. 4.3). Three different materials were used (see Petrucci and Zuccarello [164]). The load (spectrum) was always normalized to the value defined in Table 10.1 for the specific material data used.

TABLE 10.1 **Material fatigue parameters [164], as denoted in Eq. (4.1) and used for time-to-failure estimate calculation.**

	C [MPa^k]	k	S_u [MPa]	RMS [MPa]
Steel	$1.934 \cdot 10^{13}$	3.624	725	5
Aluminium	$1.08 \cdot 10^{22}$	7.300	446	10
Spring steel	$4.9 \cdot 10^{40}$	11.760	1850	100

The rainflow algorithm in the time domain provides the information about amplitudes and averages of loading cycles. Due to the lack of information about the cycle averages in the frequency domain methods, the evaluation of the accuracy has not been attempted for the cycle averages.

The results of the following spectral moments methods were compared for accuracy to the rainflow counting method (see Sec. 4.2.2):

- **NB** Narrowband (4.10),
- **WL** Wirsching–Light (4.12),
- **AL** $\alpha_{0.75}$ (4.16),
- **GM** Gao–Moan (4.17),
- **DK** Dirlik (4.28),
- **ZB1** Zhao–Baker 1 (4.36),
- **ZB2** Zhao–Baker 2 (4.36) and (4.33)–(4.35),
- **TB1** Tovo–Benasciutti 1 (4.42), (4.43),
- **TB2** Tovo–Benasciutti 2 (4.42), (4.44).

Furthermore, the consistency of estimates for the respective method was checked. The best performing methods should show good accuracy regardless of spectra group and spectrum features. The results are given in Tables 10.2–10.4. The performance of each method is described by the percentage of results, which were within the 5%, 10%, 20%, and 50% relative error margin. For example, the improved Zhao–Baker method (Sec. 4.3.6) gave estimates with relative error of 5% or better on 65% of simulated spectra overall.

TABLE 10.2 The percentage of the results within the bounds of relative errors, for S-N slope $k = 3.624$.

	Relative error			
	< 0.05	< 0.1	< 0.2	< 0.5
NB	0	22	76	100
WL	36	80	100	100
AL	51	100	100	100
GM	0	0	39	93
DK	86	100	100	100
ZB1	42	91	100	100
ZB2	65	96	100	100
TB1	0	26	80	100
TB2	100	100	100	100
	% of results			

TABLE 10.3 The percentage of the results within the bounds of relative errors, for S-N slope $k = 7.3$.

	Relative error			
	< 0.05	< 0.1	< 0.2	< 0.5
NB	0	0	16	100
WL	43	69	96	80
AL	13	25	59	100
GM	0	0	0	93
DK	15	37	83	100
ZB1	0	25	67	100
ZB2	20	25	59	100
TB1	0	4	20	100
TB2	32	73	100	100
	% of results			

10.3.1 Detailed comparison

Four models, Tovo–Benasciutti, Dirlik, and both Zhao–Baker, show exceptional accuracy compared to other spectral methods and are analyzed in greater detail. The performance of these methods can be seen for each group and material in Figs. 10.7–10.9.

In the multimodal group (Fig. 10.2) the estimates for $k = 3.624$ were most accurate in the case of the two- and three-peak spectrum. The times-to-failures were overestimated for the single-peak spectrum, whereas for more peaks, they very conservative. The relative errors are generally small for the considered coefficient k but get progressively bigger and more conservative for higher values

TABLE 10.4 The percentage of the results within the bounds of relative errors, for the S-N slope $k = 11.76$.

	relative error			
	< 0.05	< 0.1	< 0.2	< 0.5
NB	0	0	0	82
WL	33	59	91	100
AL	0	7	13	95
GM	0	0	0	7
DK	0	7	17	95
ZB1	0	0	12	88
ZB2	0	7	13	68
TB1	0	0	0	82
TB2	11	24	62	100
	% of results			

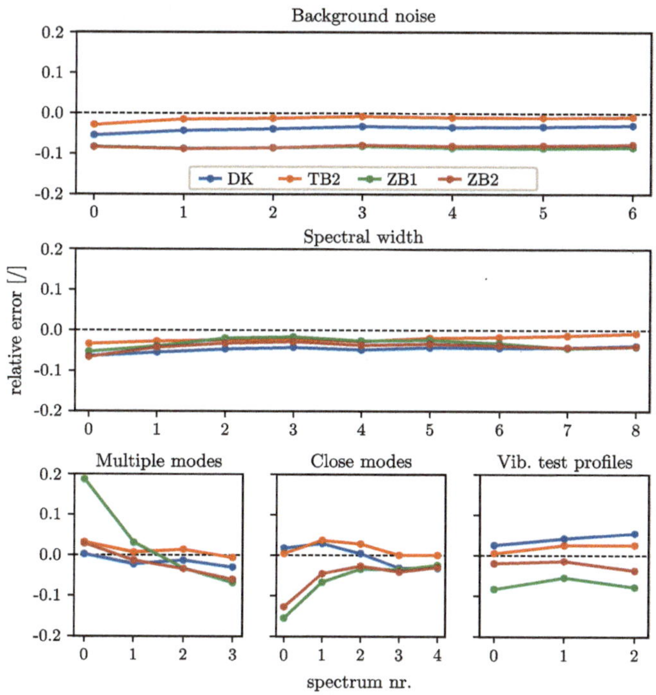

FIGURE 10.7 Relative errors of Tovo–Benasciutti, Dirlik, and both Zhao–Baker methods at $k = 3.624$.

of the S-N slope coefficient k. The TB2 method gives the best accuracy overall, whereas the DK method is the only one that gives conservative estimates throughout the multimodal spectral group.

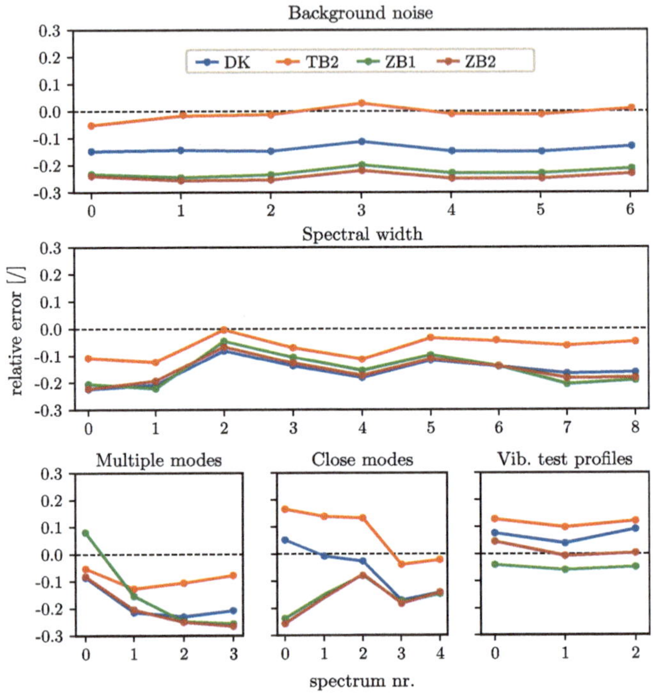

FIGURE 10.8 Relative errors of Tovo–Benasciutti, Dirlik, and both Zhao–Baker methods at $k = 7.3$.

The TB2 method gives exceptional accuracy in the background noise level group of spectra (Fig. 10.3), even though it does overestimate in some cases. The results are very accurate for $k = 3.624$, which is also true for the remaining groups of spectra. The DK method is strictly conservative within this group, but the TB2 is recommended due to its superior accuracy.

The estimates are again very accurate for $k = 3.624$ in the spectral width group (Fig. 10.4) for all four methods. Some inconsistencies are uncovered when applying higher values of k. The best performing method in this group is the TB2, giving an accurate and always conservative estimate.

Close-modes group spectra (Fig. 10.5) show that all four methods are inconsistent for the case of well-separated peaks. DK and TB2, the methods that show the best accuracy in this group, can also overestimate time-to-failure when the distance between the peaks is large. The ZB1 and ZB2 methods are always conservative within this group of spectra but also have lower accuracy than the competing methods, especially for high values of k.

The ZB1 and ZB2 are the most accurate methods across the typical accelerated vibration profiles (Fig. 10.6). The ZB1 is more accurate but can also slightly overestimate the time-to-failure, whereas the ZB2 is strictly conserva-

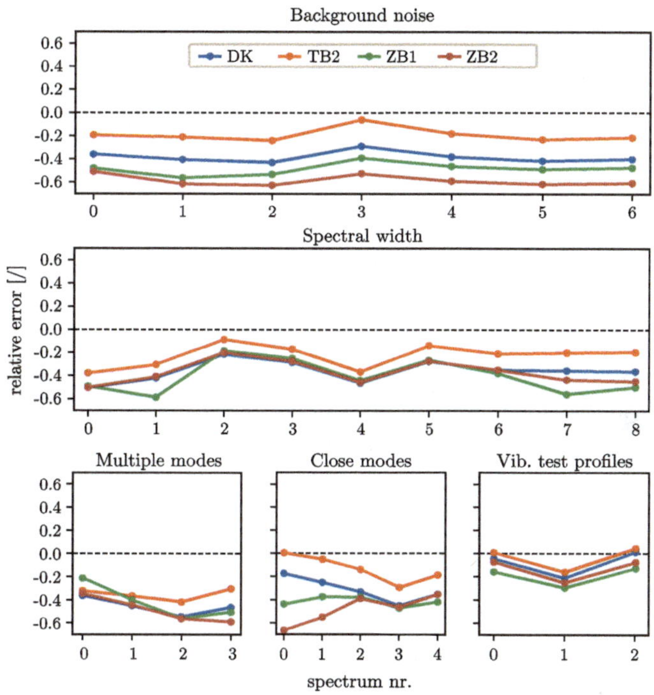

FIGURE 10.9 Relative errors of Tovo–Benasciutti, Dirlik, and both Zhao–Baker methods at $k = 11.76$.

tive. The DK and ZB2 methods are more accurate only for the highest value of $k = 11.760$.

10.4 Choosing the optimal spectral moments method

Taking into account all 28 spectra, the best performing method overall was the Tovo–Benasciutti method (TB2). This hints that the choice of method would usually be either the Dirlik method (already well established) or the Tovo–Benasciutti method for a general equivalent stress response PSD.

The best performing methods in the study, the Dirlik, Zhao–Baker, and Tovo–Benasciutti, give accurate and consistent estimates for a low value of the coefficient $k \approx 3$ (Fig. 10.7). For a steeper S-N curve slope when $k = 7.30$ or $k = 11.76$, the estimates get progressively less accurate. The small inconsistencies are amplified, and the reliability of the methods is poor for $k = 11.76$ (Fig. 10.9). This finding about the importance of effect of the material data on the final time-to-failure estimate is consistent across multiple studies [60,160]. It could be even anticipated by looking at Eq. (4.9), where k is the stress exponent and thus amplifies the existing stress estimate error. Estimates obtained for a very steep S-N curve should always be carefully evaluated.

Chapter 11

Multiaxial vibration fatigue

Vibration fatigue analysis is concerned with the cases where a structure has a pronounced dynamic response over a broadband frequency range, where it is randomly excited. Such cases represent a somewhat specific subcategory of fatigue, and the collected multiaxial criteria (Sec. 5.3) may be more or less applicable. To evaluate the accuracy and performance of criteria for a general vibration fatigue example, an experiment has been devised by Mršnik et al. [83], in which two uncorrelated (2.30) random vibration sources (electrodynamical shakers) were used to induce failure in a Y-shaped structure, shown in Fig. 11.1. The loads in material are a consequence of two resonant modes, each excited by one of the shakers. The combination of excited modes creates a complex multiaxial state within a material point and is thus an appropriate benchmark for the presented criteria.

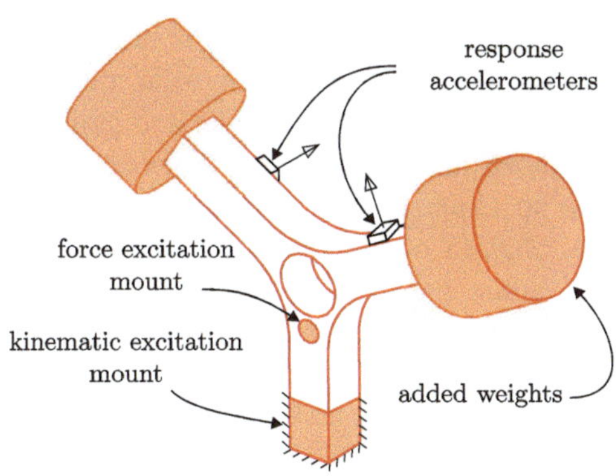

FIGURE 11.1 Schematics of the Y-sample with sensor positions and fixture mounts.

11.1 Experimental model

The Y-sample is made from cast aluminum machined to the specified geometry. Two masses are mounted at each arm, which help to tune the resonant frequencies for the desired range. A threaded hole is drilled near the center of the structure that makes for a robust force excitation mount; see Fig. 11.2. Two

Vibration Fatigue by Spectral Methods. https://doi.org/10.1016/B978-0-12-822190-7.00019-0

accelerometers, one on each arm, are used to monitor the response in real time and to continuously determine the modal frequencies and damping.

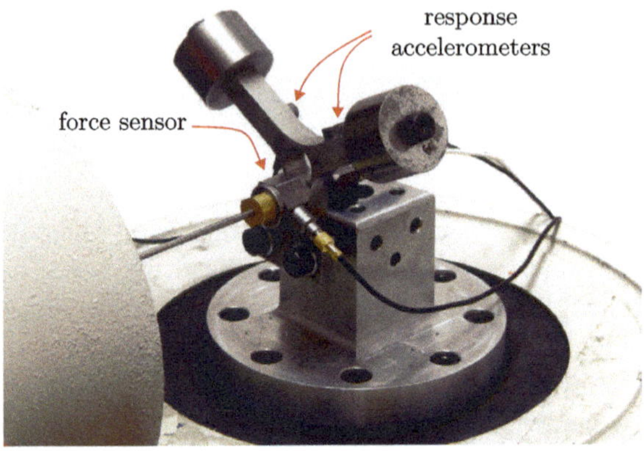

FIGURE 11.2 Experimental setup.

The excitation, as demonstrated in Fig. 11.4, is applied from two perpendicular directions. The broadband random kinematic excitation in the range of 380–480 Hz is input through a shaker adapter to which the Y-specimen is fixed. The vibrations are controlled in a closed loop by means of the reference accelerometer mounted on the fixture.

The broadband random force excitation in the range of 290–390 Hz is applied through the stinger and force transducer from the side. The vibration is controlled in an open loop by setting the signal and gain on the shaker input. This gives some additional uncertainty on the force excitation part of the load, as the force felt by the specimen depends on its dynamic properties.

The combinations of *RMS* of force and kinematic loads are varied so that the change in time-to-failure and the accuracy of the numerical analysis can be evaluated thoroughly. Ten combinations for ten specimens that were brought to failure in the course of experiments are presented in Fig. 11.3 and Table 11.1. These are the values that were experimentally measured in the beginning of each experiment and were subsequently used in simulations to get the best possible accuracy of the numerical time-to-failure estimates.

The dynamic response of the specimen changes as the damage within material accumulates. The bandwidth of the excitation signals guarantees that the natural frequencies are within the excitation range for the complete duration of the test until the specimen fails. However, this phenomenon (the falling of the natural frequency) is the reason for the discrepancies between the numerical and experimental models. To make the vibration fatigue analysis, which uses a linear response model, feasible, we update the numerical model before the experiment starts and assume that the response does not change.

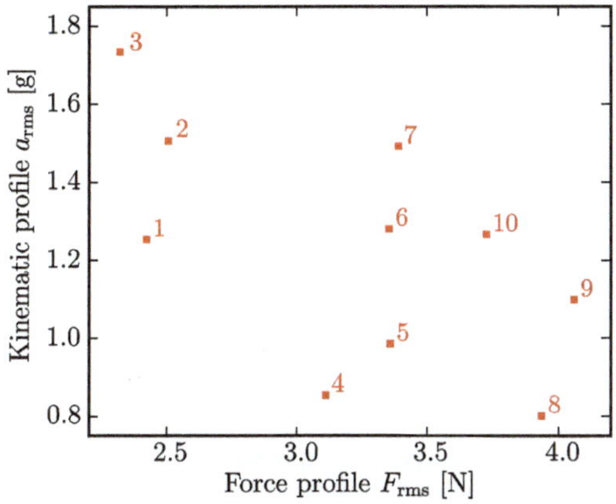

FIGURE 11.3 The force and kinematic load combinations as measured in the beginning of each experiment.

TABLE 11.1 The force and kinematic load combination values as measured in the beginning of each experiment.

Load case	F_{rms} [N]	a_{rms} [g]
1	2.42	1.25
2	2.51	1.51
3	2.32	1.73
4	3.11	0.86
5	3.36	0.99
6	3.36	1.28
7	3.39	1.49
8	3.94	0.80
9	4.06	1.10
10	3.73	1.27

11.2 Numerical model

We prepared a linear numeric model using the finite element method (FEM) and performed a frequency analysis yielding first eight (up to 3285 Hz) eigenvectors and eigenvalues of the model. We calculated the random response for modal shapes that were to be excited, the force-excited mode I at the calculated 355 Hz natural frequency and the kinematically excited mode II at 472 Hz. The frequency-domain stress responses were obtained for the critical area of the model using the modal superposition (1.60) and structural modification approaches [6]. The modal shapes are shown in Figs. 11.5 and 11.6.

The accuracy of the model for one measured sample is shown in Figs. 11.7 and 11.8. The relative error of numerical model with regard to the calculated

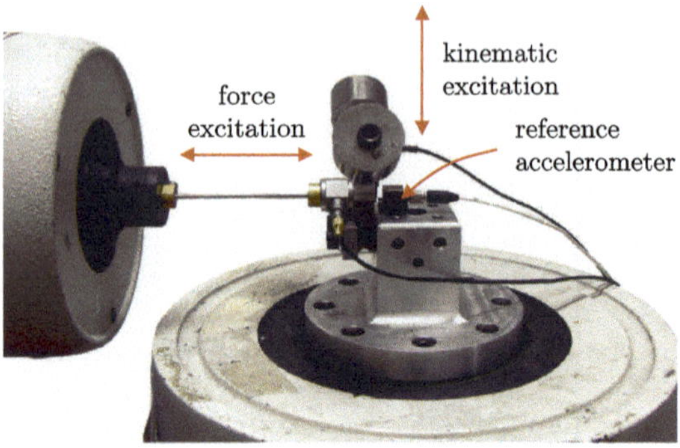

FIGURE 11.4 Excitation with broadband random vibration.

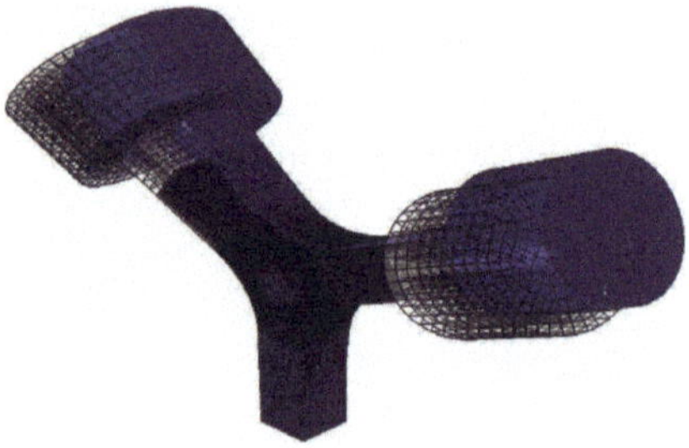

FIGURE 11.5 Modal shape I.

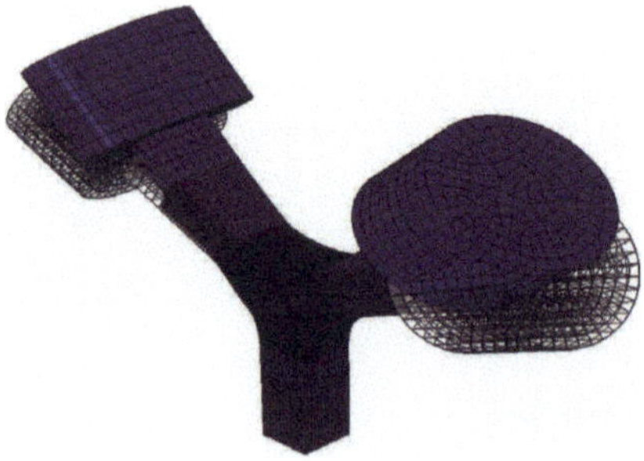

FIGURE 11.6 Modal shape II.

natural frequency is 2.5% for mode I and 0.9% for mode II. Such responses were simulated for each specimen by updating the numerical model with experimentally obtained modal damping ratios. The ratios were identified using the circle-fit [4] method and the hysteretic damping model [24]. The scatter is reduced in estimates by means of model updating, as the damping ratios generally vary slightly across specimens.

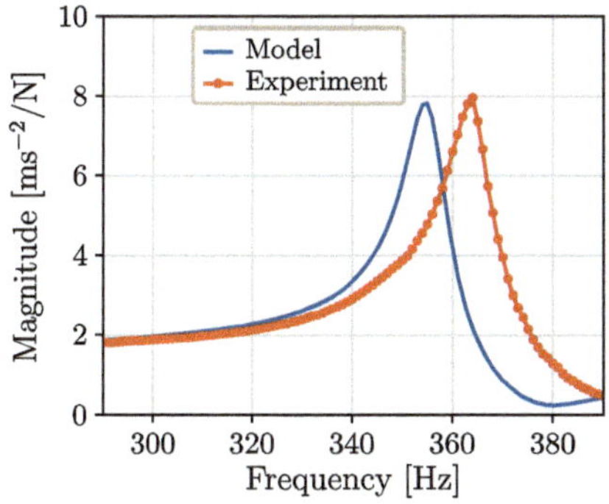

FIGURE 11.7 Comparison of measured and simulated FRF of the specimen in the frequency range and direction of random force excitation.

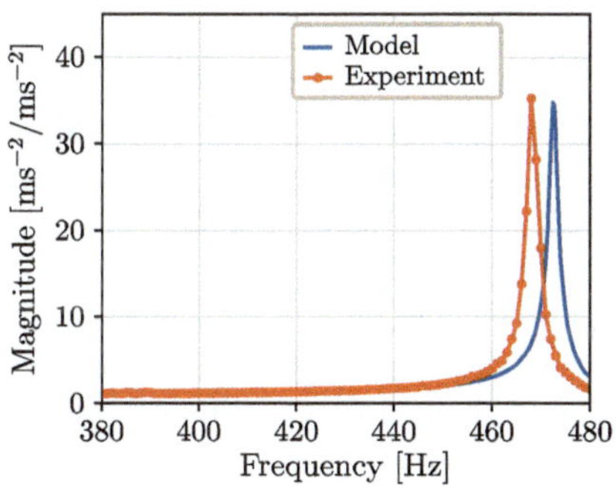

FIGURE 11.8 Comparison of measured and simulated FRF of the specimen in the frequency range and direction of random kinematic excitation.

11.3 Failure detection

It is necessary to define the point of failure for a consistent comparison of specimen failures and numerical time-to-failure estimates. The strategy of loading a sample until it breaks would be ambiguous in a way, considering that the numerical models are constant and the dynamics of the specimen changes significantly in the process of breaking down. It is more common to specify failure when a crack becomes visible, usually with length of order 1 mm or less, which also corresponds to the standard stress-life analysis procedure.

Here we use a damage model that correlates the accumulated damage with a drop in the natural frequency, which can be determined with a good accuracy by continuously measuring the kinematic response with an accelerometer.

The FBDD (Frequency-based damage detection) method by Kim et al. [165] suggests that the damage α_j at position j and the sensitivity F_{ij} of the specific modal shape i at position j are proportional to the relative change in natural frequency Z_i:

$$\sum_{j=1}^{\mathrm{NE}} F_{ij}\,\alpha_j = Z_i, \tag{11.1}$$

where $Z_i = \Delta\omega_i^2/\omega_{i,0}^2$, and NE is the number of elements. For this comparison, we assume that the sensitivity is the same for each element and that the relative change $Z_i = 5\%$ in the natural frequency $\omega_{i,0}$ defines the specimen failure equally for both modal shapes ($Z_i = Z_1 = Z_2$). Based on Z_i, the absolute fall $\Delta\omega_i$ of natural frequency $\omega_{i,0}$ can be written as

$$\Delta\omega_i = \omega_{i,0} - \sqrt{\omega_{i,0}^2 - 0.05\,\omega_{i,0}^2}, \tag{11.2}$$

where $\omega_{i,0}$ is determined by means of experiment for both modal shapes ($i = 1, 2$) for each specimen just before the start of the experiment. The specimen has failed when either frequency $\omega_{i,0}$ falls by $\Delta\omega_i$.

It is necessary to mention a certain inconsistency with regard to the approach described in (11.2). Although the fall of one natural frequency is consistent across specimen, the deltas in natural frequencies of modes I and II are not necessarily consistent for a single specimen in terms of accumulated damage. This means that, for example, a one percent fall of the mode I natural frequency does not represent a proportional damage equal to one percent fall of mode II natural frequency. It is also not clear how the damage due to mode I changes the natural frequency of mode II and vice versa.

11.4 Time-to-failure

The random stress response calculated using updated numerical model was evaluated, and each presented multiaxial criterion (Sec. 5.3) was then applied to obtain the time-to-failure for each specimen. Material parameters are deduced that minimize the difference between the numerical and experimental results, separately for every multiaxial criterion formulation across all 10 specimens. In the first pass of the optimization, standard material parameters for steel (Table 10.1) were used to obtain the critical point on the structure. In the second pass the L-BFGS-B [166] optimization method was used to get the optimal coefficients k and C of the S-N curve by searching the minimum of the function $\mathcal{L}(k, C)$, which represents the sum of squares of differences between numerical estimates $t_{i,\mathrm{num}}$ and experimental times $t_{i,\mathrm{exp}}$:

$$\mathcal{L}(k, C) = \sum_i \left(t_{i,\mathrm{exp}} - t_{i,\mathrm{num}}\right)^2 \tag{11.3}$$

for all specimens i. Only the S-N curve parameters were determined in this way. For the case of the C-S criterion, the σ_{af} and τ_{af} values, which could not be obtained for the AlSi7Cu3 alloy, the substituted values $\sigma_{af} = 161$ MPa and $\tau_{af} = 97$ MPa were taken from [167] for AlCuMg1.

The times-to-failure ranged from 20 min to 135 min, depending on the specimen and the load severity. Between 9×10^5 and 6×10^6 loading cycles occurred during the experiment, which justifies use of the stress model. The S-N curve coefficients obtained through the optimization method are listed in Table 11.2. These results pertain to the component, which is the Y-specimen, and cannot be directly compared to the standard experiments on smooth tubular specimens.

TABLE 11.2 The S-N curve coefficient that gave best time-to-failure estimates for respective multiaxial criterion.

Multiaxial criterion	C [MPa^k]	k
Max. shear stress	4.71×10^{22}	1.92
Max. normal stress	5.37×10^{23}	2.18
EVMS	4.99×10^{21}	1.92
Projection-by-projection	2.23×10^{18}	1.71
Carpinteri–Spagnoli	3.83×10^{18}	1.51
Max. normal-and-shear stress	1.11×10^{21}	1.85

The correlation plot in Fig. 11.9 shows reasonable scatter, considering the degree of uncertainty in the complete numerical analysis, from structural response calculation to damage accumulation. Most of the estimates are within the ±200% scatter band, denoted by the inner dashed line, and only one estimate that deviates for more than ±300%.

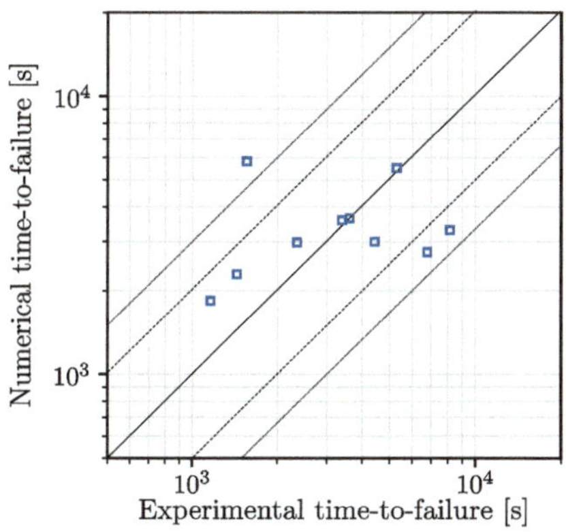

FIGURE 11.9 Correlation of numerical and experimental results for the maximum shear stress criterion.

A comparison across all the criteria is performed based on the relative difference D between the experimental and numerical time-to-failure:

$$D = \frac{t_{\text{exp}} - t_{\text{num}}}{t_{\text{exp}}}, \tag{11.4}$$

where t_{exp} is the experimentally measured time-to-failure, and t_{num} is the numerical estimate. A broader comparison in Fig. 11.10 asserts that the discrepancies between the results obtained using different multiaxial criteria are small. Furthermore, evaluating Fig. 11.10 and taking into account the load combinations from Fig. 11.3 reveals that the discrepancies are larger when the force excitation is the predominate load.

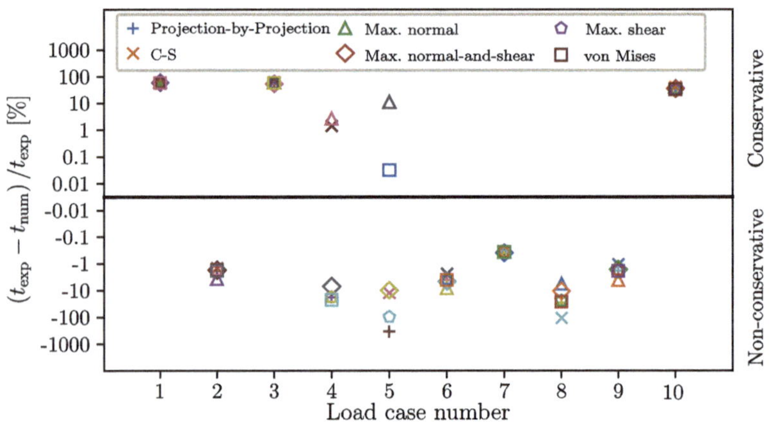

FIGURE 11.10 Correlation of numerical and experimental results for the maximum shear stress criterion.

11.5 Crack location

The comparison of multiaxial criteria with regard to the crack location estimate also shows only negligible differences for the case of vibration fatigue. This can be seen in Fig. 11.11 by comparing the damage intensity distribution calculated for the numerical model. The critical location estimates are fairly close, and discrepancies are very hard to observe visually. This conclusion also extends to other compared multiaxial criteria.

On the contrary, the experimental results show a noticeable correlation between the crack offset and the ratio $\delta = F_{\text{rms}}/a_{\text{rms}}$. The trend of crack moving together with the change in δ is evident, despite the scatter, in Fig. 11.12, depicting the samples stacked side by side in the ascending order of δ. Such a phenomenon can be the consequence of the interactions between the two modal shapes that were excited during each experiment.

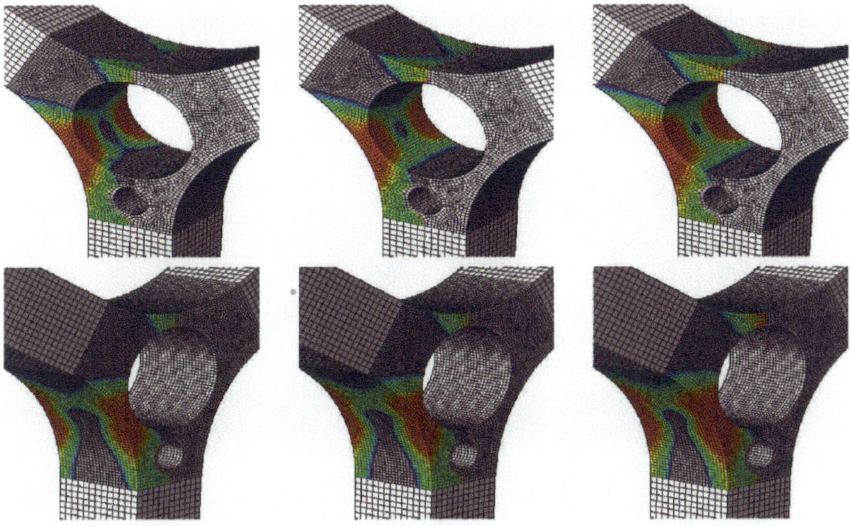

FIGURE 11.11 The damage intensity contours for the loading case 6, as calculated using (A) PbP approach, (B) C-S criterion, and (C) EVMS.

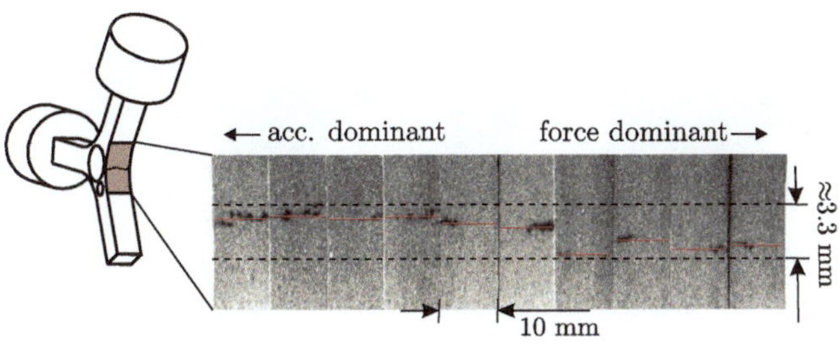

FIGURE 11.12 Progression of crack location with increase of ratio δ.

A numerical experiment is performed at this point with the aim of numerically producing a similar result. Two numerical analyses are performed, each only taking into account either force C_1 or kinematic load C_2. The critical point estimate gives two different locations for C_1 and C_2, as seen in Fig. 11.13, and then it is reasonable to assume that combining these two loads in a numerical simulation should yield a critical point somewhere on a path connecting C_1 and C_2.

The actual results are unexpected though, as the critical points for different force-kinematic load combinations are mostly fixed at a point in the vicinity of C_1 and move suddenly for the load case 3 (see Fig. 11.14). The finite-element size is 0.4 mm in the vicinity of the critical point, which is smaller than the change in the critical location (around 3 mm). Such a result can point to a possible discrepancies in the vibration fatigue analysis of a randomly excited sample. However, it should be noted that the critical point was deduced for experiment from a large crack and that the precise point of crack initiation was not determined.

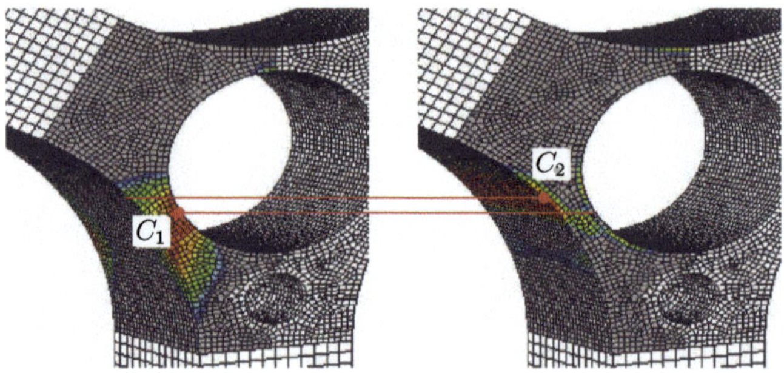

(A) (B)

FIGURE 11.13 Comparisons of failure locations: (A) force C_1 excitation and (B) kinematic C_2 excitation.

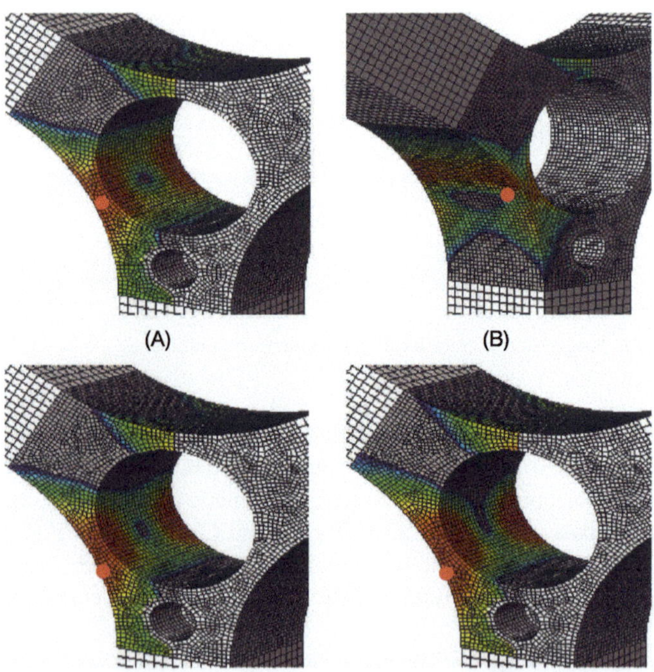

(A) (B) (C) (D)

FIGURE 11.14 Critical points calculated with the C–S multiaxial criterion for different load cases (see Table 11.1): (A) load case 2, (B) load case 3, (C) load case 6, and (D) load case 8.

Chapter 12

Vibration fatigue for nonstationary and non-Gaussian excitation

Sec. 12.1 focuses on the non-Gaussianity and the nonstationarity of random excitation signal and its influence on actual vibration fatigue life as was initially researched by Palmieri et al. [121]. Section 12.2 focuses on the nonstationarity and on how to properly identify it as was initially studied by Capponi et al. [122].

12.1 Nonstationarity vs. non-Gaussianity

For this purpose, we use a Y-shaped specimen (see Fig. 8.8), as already presented in Ch. 8. The specimen was attached to an electrodynamic shaker and excited with flat-shaped force PSD profile (Fig. 12.1) with frequency range 600–850 Hz, so that the specimen's fourth natural frequency of 775 Hz lied within the excited frequency band.

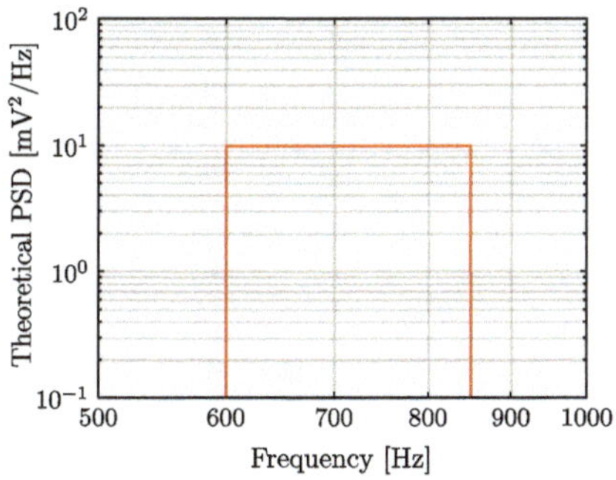

FIGURE 12.1 Flat PSD profile with amplitude level 10 mV^2/Hz.

The procedure for evaluating non-Gaussianity and nonstationarity is as follows. First, a set of specimens is excited with Gaussian stationary signals with flat PSD profiles at different amplitude levels; see Fig. 12.1. By minimizing the error between the experimental and numerical fatigue lives the material fatigue parameters C and k are obtained according to their definition in Eq. (4.1). Second, random signals with similar PSD profiles but different kurtosis (Eq. (7.3)) and nonstationarities (Table 12.1) are generated and applied to the Y-specimens.

Vibration Fatigue by Spectral Methods. https://doi.org/10.1016/B978-0-12-822190-7.00020-7

Influence of non-Gaussianity and nonstationarity is then studied by comparing actual and predicted fatigue lives for given excitation signal types.

TABLE 12.1 Excitation signal types.

Nr.	Signal type	k_u
1.	Stationary Gaussian	2.96
2.	Stationary non-Gaussian	7.36
3.	Stationary non-Gaussian	5.43
4.	Nonstationary non-Gaussian	7.08

Four Y-shaped specimens were excited with random stationary Gaussian signal until fatigue failure occurred. The obtained fatigue lives are presented by blue circles in Fig. 12.2. To obtain fatigue material parameters, a numeric model was built and validated (Fig. 12.3). The following function for numerical minimization of numerical error was proposed:

$$\Delta T(k, C) = \sum_i (\log_{10}(T_{\text{act},i}) - \log_{10}(T_{\text{est},i}(k, C)))^2, \tag{12.1}$$

where i denotes a single test specimen, T_{act} is the measured fatigue life, and T_{est} is the estimated fatigue life of the specimen based on the numerical model. The minimization method showed strong convergence. Using the identified fatigue parameters, the Basquin equation (Eq. (4.2)) can be written as

$$\sigma = 987.5 \cdot N^{-0.169}. \tag{12.2}$$

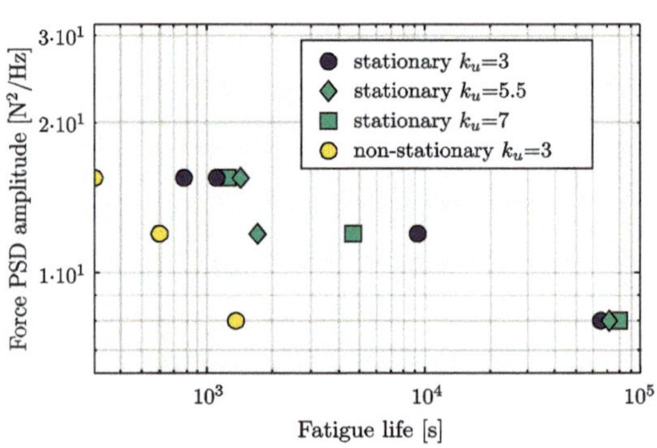

FIGURE 12.2 Experimental fatigue lives for non-Gaussian and nonstationary excitation.

Once fatigue parameters were known, a number of samples were excited with signals, where non-Gaussianity and nonstationarity were combined according to Table 12.1. Time signals for signal types 2 and 4 with identical PSD profiles are presented in Figs. 12.4 and 12.5. The excitation force PSD level

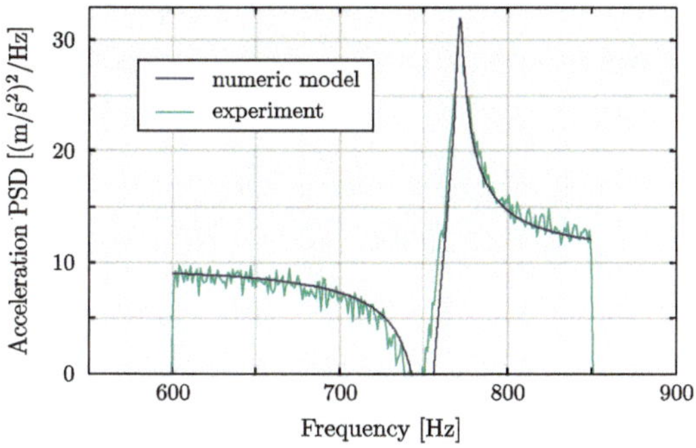

FIGURE 12.3 Numerically and experimentally obtained response acceleration PSD of tested Y-specimen.

was beforehand determined by measuring acceleration response of shaker armature without Y-specimen [121]. A nonstationary signal was generated using amplitude modulation of stationary signal with beta-distribution signal [121, 122].

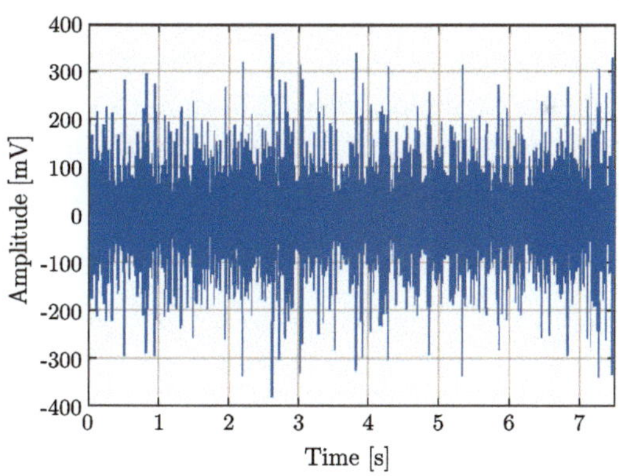

FIGURE 12.4 Stationary non-Gaussian signal with $k_u = 7$.

A total of nine Y-shaped specimens were tested under non-Gaussian excitation, three for each signal type 2–4. To evaluate the influence of the kurtosis k_u on the fatigue life of tested specimens, the fatigue parameters obtained under the Gaussian condition are also used for numerical estimation of the fatigue life for the case of non-Gaussian excitation. To investigate when it is justified to apply Gaussian-based counting methods (see Sec. 4.3) to nonstationary excitation, the Tovo–Benasciutti counting method (Eq. (4.42)) was also used in the case of non-Gaussian excitations.

A comparison between the experimental and calculated fatigue lives is shown in Fig. 12.6. It is noticeable that the stationary non-Gaussian excitation differs only slightly if compared to the calculated fatigue life. In contrast

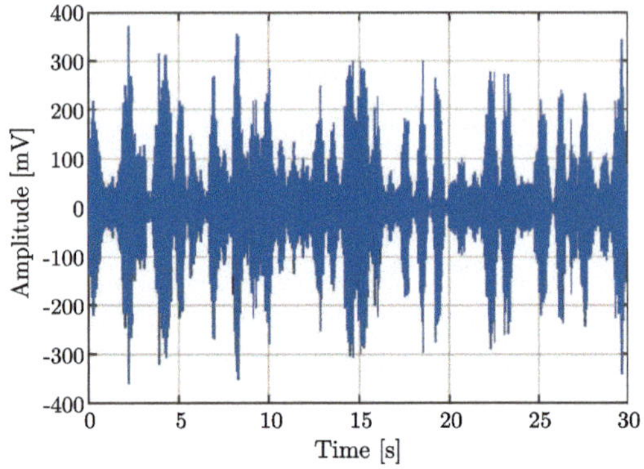

FIGURE 12.5 Nonstationary non-Gaussian signal with $k_u = 7$.

to this case, for the case of nonstationary non-Gaussian excitation, the comparison shows how the difference between the calculated fatigue lives and the actual fatigue lives becomes significant. From this the use of standard frequency-counting methods [66] available in the literature gives reliable results comparable to reality, whereas in the case of strongly nonstationary signals the use of the same counting methods supplies uncorrected results. Also, Fig. 12.2 shows the relation between the excitation force PSD amplitude and the experimental fatigue life; again, the nonstationary experiments clearly differ from the stationary experiments.

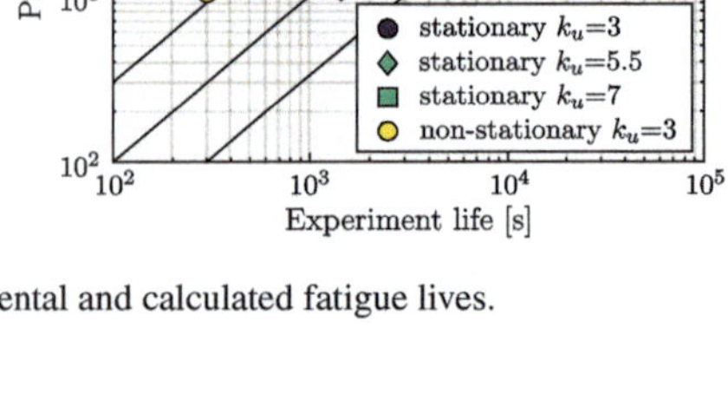

FIGURE 12.6 Experimental and calculated fatigue lives.

12.2 Nonstationarity index

The following section focuses on quantification of nonstationarities in excitation signal and on identification of nonstationary signals that result in reduced vibration fatigue life.

12.2.1 Signal generation

First, a set of time signals with different nonstationarity rates was obtained. For the bulk of physical phenomena, the nonstationarity is exhibited as the time variance of a signal power; the fluctuations in the signal frequency content are less commonly observed. Therefore beta-distribution amplitude modulation of a stationary signal with flat-shaped PSD profile was initially performed with a carrier wave, as proposed in Sec. 12.1 (Fig. 12.5) and described in detail in [131] and [121].

In the next step the initial carrier wave with time duration of 30 min was squeezed by a factor of 2 and again multiplied with flat-profiled stationary signal. This signal was denoted as SQ-2, whereas the amplitude-modulated signal with original (nonsqueezed carrier wave) was denoted as SQ-1. In a similar manner, additional five nonstationary signals were generated for different squeezing factors of 4, 10, 50, 500, and 10000. After squeezing, each signal was repeated to reach the total time length of 30 min. All the time, signals had the same PSD, the same time length, the same kurtosis, but different levels of nonstationarity. Squeezed signals SQ-10, SQ-50, SQ-500, and SQ-10000, which were measured on an electrodynamic shaker, are presented in Fig. 12.7.

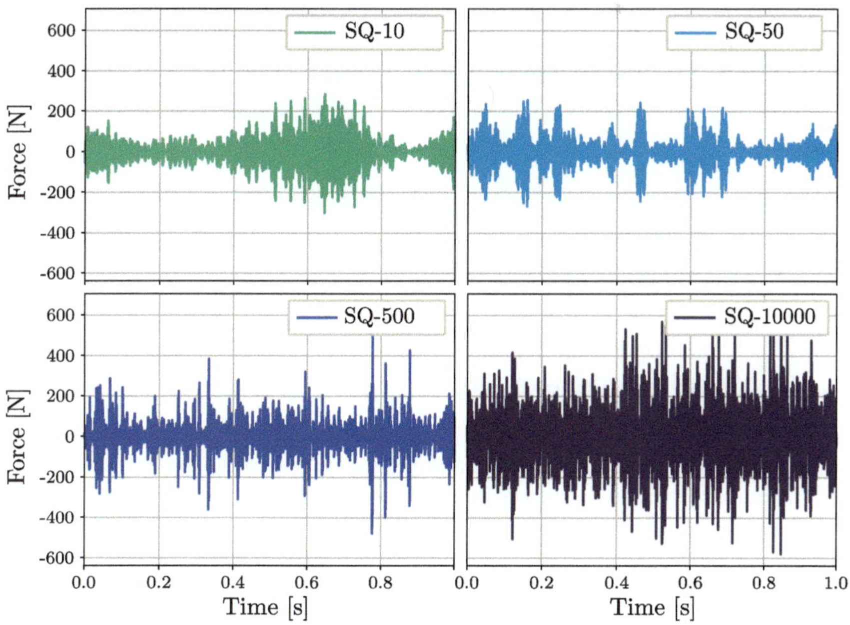

FIGURE 12.7 Time history of measured excitation signals SQ-10, SQ-50, SQ-500, and SQ-10000 with PSD level of 13.5 N^2/Hz.

12.2.2 Signal evaluation with run-test method

Here the signals later used for Y-specimen excitation and with PSD level of 13.5 N^2/Hz are evaluated with two-sided run-test method, presented in Sec. 7.2 and Fig. 7.1. As is known [73], the run-test method is sensitive to the width of the windows to which the signal is divided. Consequently, all eight signals (seven nonstationary and one de facto stationary) were evaluated for different time lengths of dividing windows. In total, seven time-window widths were applied to the run-test, ranging from 0.005 to 1 s. In Table 12.2 the resulting nonstationarity indexes γ are given. Moreover, the influence of the window width is graphically presented in Fig. 12.8. By inspecting Fig. 12.8 we observe a significant difference between low squeezing and high squeezing; at a window width of 0.04 s the signals SQ-1 to SQ-4 were identified as nonstationary, whereas the SQ-500, SQ-10000, and also de facto stationary signal were identified as stationary. Similar results were found for other RMS values of the excitation signal.

TABLE 12.2 Nonstationarity index γ for two-sided method: results in bold identify signal as stationary.

Signal type	Window width						
	0.005 s	*0.01 s*	*0.0125 s*	*0.02 s*	*0.04 s*	*0.1 s*	*1 s*
	Nonstationarity index γ [%]						
Stationary	96	**99**	**100**	**101**	**102**	**99**	72
SQ-10000	96	**101**	**101**	**101**	**102**	**100**	**101**
SQ-500	95	97	**99**	**99**	**102**	105	**90**
SQ-50	73	86	92	98	95	96	114
SQ-10	65	62	61	61	72	**96**	**109**
SQ-4	64	60	57	50	46	67	**105**
SQ-2	65	60	58	50	40	39	**104**
SQ-1	66	62	60	53	40	31	**96**

12.2.3 Vibration fatigue testing

Here the de facto stationary and squeezed signals SQ-1–SQ-10000 were applied to the Y-specimen. According to a preliminary analysis of the excitation signals, the nonstationarity index γ shows that the signals SQ-500 and SQ-10000 can be considered as stationary and the signals SQ-1, SQ-2, and SQ-4 can be considered as nonstationary. This finding was now tested against the actual vibration fatigue life.

For experimental comprehensiveness, the time signals were applied to the Y-specimen at four different force PSD levels, as shown in Fig. 12.9. For each of the 23 combination pairs of excitation PSD level and signal type, two samples were tested. In total, 42 samples were tested.

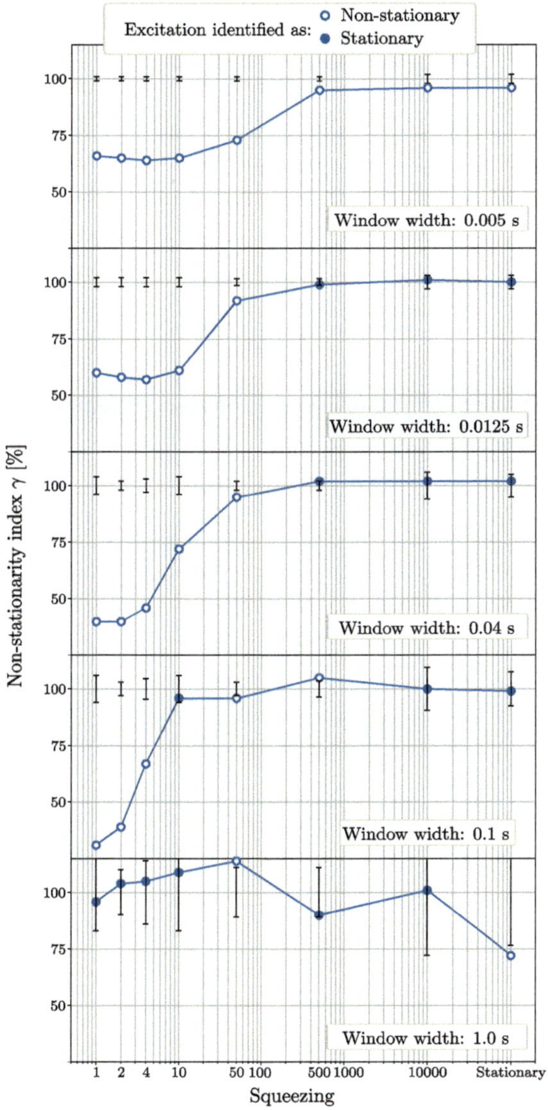

FIGURE 12.8 Influence of window width on the nonstationarity index γ of measured excitation signal for two-sided run-test method with denoted confidence intervals.

The fatigue life for all tested specimens is given in Table 12.3. It is clear that squeezing the carrier wave and thus changing the nonstationarity rate significantly changes the fatigue life; see Fig. 12.10. From Fig. 12.10 it is clear that for the SQ-4, the fatigue life is significantly shorter than for the de facto stationary signal, whereas SQ-50 is in between.

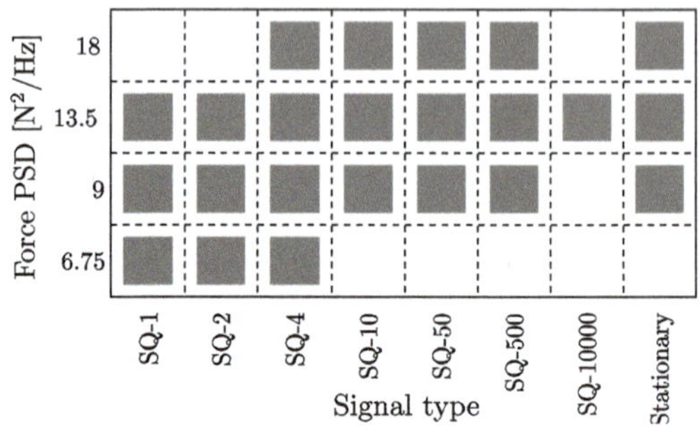

FIGURE 12.9 Experimentally tested combinations of signal types and excitation levels.

TABLE 12.3 Experimental fatigue lives [s] of tested Y-specimens.

Force PSD level [N^2/Hz]	Signal type							
	SQ-1	*SQ-2*	*SQ-4*	*SQ-10*	*SQ-50*	*SQ-500*	*SQ-10000*	*Stationary*
18	/	/	643 341	598 497	999 1069	1624 2793	/	3157
13.5	961 857	871 907	737 899	1175 1333	1120 5234	4069 6411	6299 6309	4428
9	2364 2994	2853 2391	3279 2899	3177 3824	3722 5894	>12600	/	>12600
6.75	5112 4865	2737 4405	4429 5435	/	/	/	/	/

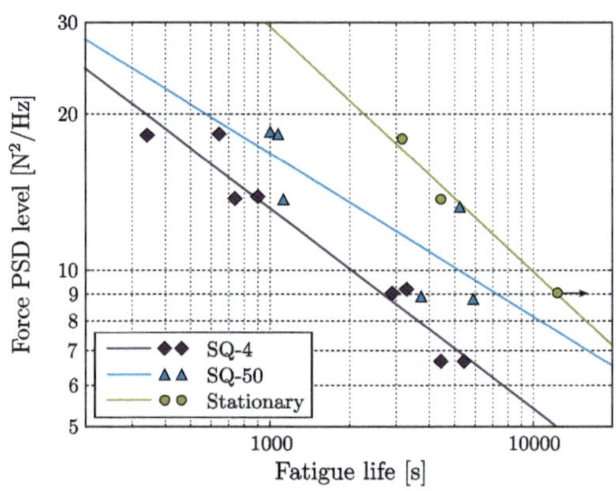

FIGURE 12.10 Experimental fatigue life for the representative excitation signal types: SQ-4, SQ-50, and de facto stationary.

Due to the considerable influence of the nonstationarity rate on the fatigue life, only tests with an excitation force PSD level of 13.5 N^2/Hz were possible on the complete set of generated nonstationary signals SQ-1–SQ-10000 and on the de facto stationary signal; see Table 12.3. To better understand Fig. 12.10, a detailed analysis of excitation with PSD level 13.5 N^2/Hz is shown in Fig. 12.11: one axis shows the fatigue life, and the other axis shows the nonstationarity index. It was previously observed that the excitation signals from SQ-1 to SQ-4 can be considered as nonstationary, whereas the signals SQ-500 and above can be considered as stationary. As all the time signals had the same PSD, the same time length, the same kurtosis, but different rates of nonstationarity, we can conclude that nonstationarity results in significant differences in fatigue life. The difference between nonstationary and stationary conditions is approximately 5-fold.

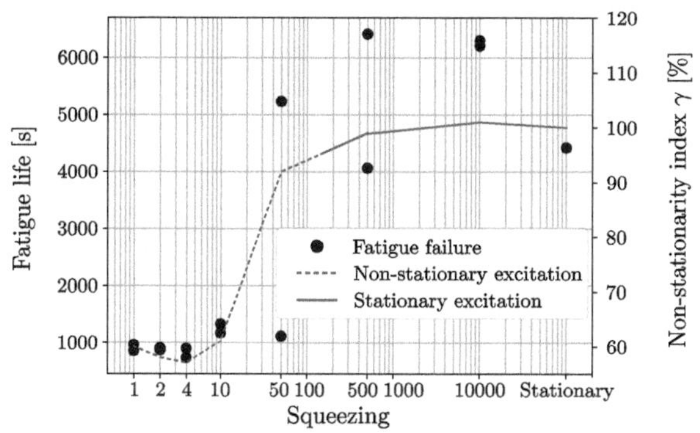

FIGURE 12.11 Comparison of experimental fatigue lives for excitation PSD level of 13.5 N^2/Hz and the nonstationarity index γ of excitation signal for window width of 0.0125 s using the two-sided run-test method.

Fig. 12.12 shows the fatigue life normed to the fatigue life during stationary excitation. From the results of all 42 tested samples it is clear that the excitation identified as nonstationary resulted in a reduced fatigue life to approximately 20% of fatigue life under de facto stationary excitation. Signals that are neither clearly stationary nor nonstationary have a fatigue life between the two groups. Based on the presented results, it is reasonable to conclude that the nonstationary index γ provides reliable estimation to distinguish between nonstationary excitation signals that do or do not reduce vibration fatigue life when compared to stationary excitation.

According to presented results, the remaining question is which the window width is appropriate to reliably identify the nonstationarity. According to Rizzi et al. [130], the optimal window width is expected to be related to the period of the system impulse response, that is, the time in which the response amplitude reduces to 10% of the initial value [131]; in the case of the Y-specimen the period of the impulse response was experimentally determined to be in the range up to 0.2 s. With regard to the results shown in Fig. 12.8, 0.2 s would result in wrong identification of nonstationarity. The drop of impact response to 10% of the initial value therefore seems to result in a too wide time window; from the obtained experimental data we see that a drop to 80% of the initial value at impact excitation corresponds to a more reasonable ≈ 0.025 s window width.

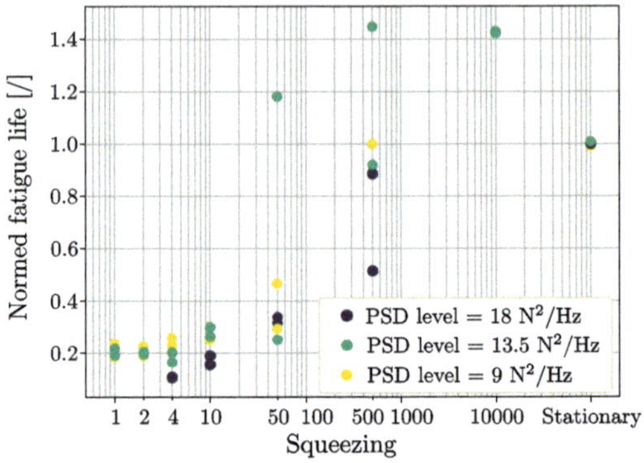

FIGURE 12.12 Fatigue lives of Y-specimens, normed to fatigue life under de facto stationary excitation signal.

Appendix A

Pseudoinverse of a matrix

For a case of an overdetermined system of equations

$$\mathbf{A}_{m\times n}\,\mathbf{x}_{n\times 1} = \mathbf{b}_{m\times 1}, \tag{A.1}$$

where $m > n$, the solution $\mathbf{x}$ cannot be determined with a classical inverse. To obtain the solution, both sides are multiplied from the left by the transposed matrix $\mathbf{A}^T$:

$$\mathbf{A}^T\,\mathbf{A}\,\mathbf{x} = \mathbf{A}^T\,\mathbf{b}. \tag{A.2}$$

The product $\mathbf{A}^T\,\mathbf{A}$ is an $n \times n$ square matrix, and an inverse can be determined allowing $\mathbf{x}$ to be expressed as

$$\mathbf{x} = \left(\mathbf{A}^T\,\mathbf{A}\right)^{-1}\mathbf{A}^T\,\mathbf{b}. \tag{A.3}$$

Therefore the pseudoinverse [168] of a matrix $\mathbf{A}$, denoted as $\mathbf{A}^\dagger$, equals

$$\mathbf{A}^\dagger = \left(\mathbf{A}^T\,\mathbf{A}\right)^{-1}\mathbf{A}^T. \tag{A.4}$$

The pseudoinverse produces a least-squares solution [168].

Bibliography

[1] Roy R. Craig, Andrew J. Kurdila, Fundamentals of Structural Dynamics, 2nd edition, John Wiley & Sons, 2011.

[2] Michel Géradin, Daniel J. Rixen, Mechanical Vibrations: Theory and Application to Structural Dynamics, 3rd edition, John Wiley & Sons, 2015.

[3] Singiresu S. Rao, Mechanical Vibrations, 5th edition, Prentice Hall, 2010.

[4] Nuno M.M. Maia, Julio.M.M. Silva, Theoretical and Experimental Modal Analysis, Research Studies Press, 1997.

[5] Daniel C. Cole, Eytan Barouch, Uwe Hollerbach, Steven A. Orszag, Derivation and simulation of higher numerical aperture scalar aerial images, Japanese Journal of Applied Physics 31 (12S) (1992) 4110–4119.

[6] Martin Česnik, Janko Slavič, Primož Čermelj, Miha Boltežar, Frequency-based structural modification for the case of base excitation, Journal of Sound and Vibration 332 (20) (2013) 5029–5039.

[7] Subra Suresh, Fatigue of Materials, 2nd edition, Cambridge University Press, 1998.

[8] O. Bernasconi, David J. Ewins, Modal strain/stress fields, The International Journal of Analytical and Experimental Modal Analysis 42 (1989) 68–76.

[9] James F. Doyle, Modern Experimental Stress Analysis: Completing the Solution of Partially Specified Problems, John Wiley & Sons, 2004.

[10] G. Thomas Mase, Ronald E. Smelser, George E. Mase, Continuum Mechanics for Engineers, 4th edition, Taylor & Francis, 2016.

[11] O. Bernasconi, David J. Ewins, Application of strain modal testing to real structures, in: Proceedings of the 7th International Modal Analysis Conference, vol. 2, 1989, pp. 1453–1464.

[12] Tadej Kranjc, Janko Slavič, Miha Boltežar, A comparison of strain and classic experimental modal analysis, Journal of Vibration and Control 22 (2) (2016) 371–381.

[13] Lai Hang Yam, Tin Pui Leung, Debao Li, Kemin Z. Xue, Theoretical and experimental study of modal strain analysis, Journal of Sound and Vibration 191 (2) (1996) 251–260.

[14] Tadej Kranjc, Janko Slavič, Miha Boltežar, The mass normalization of the displacement and strain mode shapes in a strain experimental modal analysis using the mass-change strategy, Journal of Sound and Vibration 332 (26) (2013) 6968–6981.

[15] Julius S. Bendat, Allan G. Piersol, Random Data: Analysis and Measurement Procedures, 4th edition, John Wiley & Sons, 2011.

[16] Kihong Shin, Joseph Hammond, Fundamentals of Signal Processing for Sound and Vibration Engineers, John Wiley & Sons, 2008.

[17] David E. Newland, An Introduction to Random Vibrations, Spectral & Wavelet Analysis, 3rd edition, Courier Corporation, 2012.

[18] Stephen O. Rice, Mathematical analysis of random noise, The Bell System Technical Journal 24 (1) (1945) 46–156.

[19] James W. Cooley, John W. Tukey, An algorithm for the machine calculation of complex Fourier series, Mathematics of Computation 19 (90) (1965) 297–301.

[20] Kenneth G. McConnell, Paulo S. Varoto, Vibration Testing: Theory and Practice, 2nd edition, John Wiley & Sons, 2008.
[21] William A. Fladung, Robert W. Rost, Application and correction of the exponential window for frequency response functions, Mechanical Systems and Signal Processing 11 (1) (1997) 23–36.
[22] Paul H. Wirsching, Mark C. Light, Fatigue under wide band random stresses, Journal of the Structural Division 106 (7) (1980) 1593–1607.
[23] Erik H. Vanmarcke, Properties of spectral moments with applications to random vibration, Journal of the Engineering Mechanics Division 98 (2) (1972) 425–446.
[24] David J. Ewins, Modal Testing: Theory, Practice, and Application, 2nd edition, Research Studies Press, 2000.
[25] Zhi-Fang Fu, Jimin He, Modal Analysis, Elsevier, 2001.
[26] Bart Peeters, Herman Van der Auweraer, Patrick Guillaume, Jan Leuridan, The polymax frequency-domain method: a new standard for modal parameter estimation? Shock and Vibration 11 (3–4) (2004) 395–409.
[27] Ward Heylen, Stefan Lammens, Paul Sas, Modal Analysis Theory and Testing, Katholieke Universiteit Leuven, 1998.
[28] Anders Brandt, Noise and Vibration Analysis: Signal Analysis and Experimental Procedures, John Wiley & Sons, 2011.
[29] Manfred R. Schroeder, Synthesis of low-peak-factor signals and binary sequences with low autocorrelation, IEEE Transactions on Information Theory 16 (1) (1970) 85–89.
[30] Ole J. Løkberg, Electronic speckle pattern interferometry, in: Olivério D.D. Soares (Ed.), Optical Metrology, Springer, 1987, pp. 542–572.
[31] David L. Brown, Randall J. Allemang, Ray Zimmerman, Maurice Mergeay, Parameter estimation techniques for modal analysis, in: 1979 Automotive Engineering Congress and Exposition, SAE International, 1979.
[32] Patrick Guillaume, Peter Verboven, Steve Vanlanduit, Frequency-domain maximum likelihood identification of modal parameters with confidence intervals, in: Proceedings of the 23rd International Conference on Noise and Vibration Engineering, ISMA 23, 1998, pp. 359–366.
[33] Patrick Guillaume, Peter Verboven, Steve Vanlanduit, Herman Van Der Auweraer, Bart Peeters, A poly-reference implementation of the least-squares complex frequency-domain estimator, in: Proceedings of IMAC 21, the International Modal Analysis Conference, vol. 3, 2003, pp. 1762–1770.
[34] Peter Verboven, Frequency-domain system identification for modal analysis, PhD thesis, Vrije Universiteit Brussel, 2002.
[35] Herman Van der Auweraer, Willem Leurs, Peter Mas, Luc Hermans, Modal parameter estimation from inconsistent data sets, in: Proceedings of IMAC-XVIII: A Conference on Structural Dynamics, 2000, pp. 763–771.
[36] Bart Cauberghe, Applied frequency-domain system identification in the field of experimental and operational modal analysis, PhD thesis, Vrije Universiteit Brussel, 2004.
[37] Yung-Li Lee, Jwo Pan, Richard Hathaway, Mark Barkey, Fatigue Testing and Analysis, Elsevier Butterworth–Heinemann, 2005.
[38] Xinzhong Chen, Analysis of crosswind fatigue of wind-excited structures with nonlinear aerodynamic damping, Engineering Structures 74 (2014) 145–156.
[39] Dilnei Schmidt, Lance Manuel, Hieu H. Nguyen, Luis Volnei Sudati Sagrillo, Edison Castro Prates de Lima, Fatigue reliability assessment for brace-column details in a semi-submersible hull, in: Proceedings of the ASME 2014 33rd International Conference on Ocean, Offshore and Arctic Engineering. Volume 4B: Structures, Safety and Reliability, 2014.
[40] Wallace Ferreira, Trenton Meehan, Valdir Cardoso, Neil W.M. Bishop, A comparative study of automotive system fatigue models processed in the time and frequency domain, SAE Technical Paper, 2016.
[41] Peter Wolfsteiner, Fatigue assessment of non-stationary random vibrations by using decomposition in gaussian portions, International Journal of Mechanical Sciences 127 (2017) 10–22.

[42] Lassaad Ben Fekih, Georges Kouroussis, Olivier Verlinden, Spectral-based fatigue assessment of ball grid arrays under aerospace vibratory environment, Key Engineering Materials 569 (2013) 425–432.
[43] Yadong Zhou, Qingguo Fei, Shaoqing Wu, Utilization of modal stress approach in random-vibration fatigue evaluation, Proceedings of the Institution of Mechanical Engineers, Part G: Journal of Aerospace Engineering 231 (14) (2017) 2603–2615.
[44] Yongyan Wang, Haibo Chen, Hongwei Zhou, A fatigue life estimation algorithm based on statistical energy analysis in high-frequency random processes, International Journal of Fatigue 83 (2016) 221–229.
[45] Bongsu Kang, Wayne Cai, Chin-An Tan, Dynamic stress analysis of battery tabs under ultrasonic welding, Journal of Manufacturing Science and Engineering 136 (4) (2014) 041011.
[46] August Wöhler, Versuche über die Festigkeit der Eisenbahnwagenachsen, Zeitschrift für Bauwesen 10 (1860) 160–161.
[47] Neil W.M. Bishop, Frank Sherratt, Finite Element Based Fatigue Calculations, NAFEMS, 2000.
[48] Darrell F. Socie, Gary B. Marquis, Multiaxial Fatigue, Society of Automotive Engineers, 2000.
[49] Heinrich Gerber, Bestimmung der zulässigen spannungen in eisen-constructionen, Zeitschrift des Bayerischen Architekten- und Ingenieur-Vereins 6 (6) (1874) 101–110.
[50] Bernard P. Haigh, Experiments on the fatigue of brasses, Journal of the Institute of Metals 18 (1917) 55–86.
[51] C.R. Söderberg, Factor of safety and working stress, Transactions of American Society of Mechanical Engineering 52 (1939) 13–28.
[52] John Goodman, Mechanics Applied to Engineering, Longmans, Green & Company, 1899.
[53] Claude Bathias, There is no infinite fatigue life in metallic materials, Fatigue & Fracture of Engineering Materials & Structures 22 (7) (1999) 559–565.
[54] Arvid Palmgren, Die Lebensdauer von Kugellagern, Zeitschrift des Vereines Deutscher Ingenieure 68 (14) (1924) 339–341.
[55] M.A. Miner, Cumulative damage in fatigue, Journal of Applied Mechanics 67 (1945) A159–A164.
[56] Klaus G. Eulitz, Karl L. Kotte, Damage accumulation - limitations and perspectives for fatigue life assessment, in: Proceedings of Materials Week 2000, Frankfurt, 2000.
[57] Ján Košút, History influence exponent in cumulative fatigue damage determined using two-step loading experiments, Fatigue & Fracture of Engineering Materials & Structures 25 (6) (2002) 575–586.
[58] A.S.T.M. International, E1049-85(2017) Standard Practices for Cycle Counting in Fatigue Analysis, Technical report, ASTM International, 2017.
[59] Masanori Matsuishi, Tatsuo Endo, Fatigue of metals subjected to varying stress, Japan Society of Mechanical Engineers 68 (2) (1968) 37–40.
[60] Benasciutti Denis, Fatigue analysis of random loadings, PhD thesis, University of Ferrara, 2004.
[61] Stephen D. Downing, Darrell F. Socie, Simple rainflow counting algorithms, International Journal of Fatigue 4 (1) (1982) 31–40.
[62] Claude Amzallag, J.P. Gerey, Jean L. Robert, Jean Bahuaud, Standardization of the rainflow counting method for fatigue analysis, International Journal of Fatigue 16 (4) (1994) 287–293.
[63] Igor Rychlik, A new definition of the rainflow cycle counting method, International Journal of Fatigue 9 (2) (1987) 119–121.
[64] Igor Rychlik, Fatigue and stochastic loads, Scandinavian Journal of Statistics 23 (4) (1996) 387–404.
[65] Gabriel Marsh, Colin Wignall, Philipp R. Thies, Nigel Barltrop, Atilla Incecik, Vengatesan Venugopal, Lars Johanning, Review and application of Rainflow residue processing techniques for accurate fatigue damage estimation, International Journal of Fatigue 82 (2016) 757–765.

[66] Matjaž Mršnik, Janko Slavič, Miha Boltežar, Frequency-domain methods for a vibration-fatigue-life estimation - application to real data, International Journal of Fatigue 47 (2013) 8–17.

[67] Denis Benasciutti, Roberto Tovo, Frequency-based analysis of random fatigue loads: models, hypotheses, reality, Materialwissenschaft und Werkstofftechnik 49 (3) (2018) 345–367.

[68] Loren D. Lutes, Shahram Sarkani, Random Vibrations: Analysis of Structural and Mechanical Systems, Elsevier Butterworth-Heinemann, 2004.

[69] Andrew Halfpenny, A frequency domain approach for fatigue life estimation from Finite Element Analysis, in: Proceedings of the International Conference on Damage Assessment of Structures (DAMAS 99), Dublin, 1999.

[70] Baran Yeter, Yordan Garbatov, Carlos Guedes Soares, Evaluation of fatigue damage model predictions for fixed offshore wind turbine support structures, International Journal of Fatigue 87 (2016) 71–80.

[71] Wangwen Zhao, Michael J. Baker, On the probability density function of rainflow stress range for stationary Gaussian processes, International Journal of Fatigue 14 (2) (1992) 121–135.

[72] Denis Benasciutti, Roberto Tovo, Spectral methods for lifetime prediction under wide-band stationary random processes, International Journal of Fatigue 27 (8) (2005) 867–877.

[73] Turan Dirlik, Application of computers in fatigue analysis, PhD thesis, University of Warwick, 1985.

[74] Matjaž Mršnik, Janko Slavič, Miha Boltežar, Vibration fatigue using modal decomposition, Mechanical Systems and Signal Processing 98 (2018) 548–556.

[75] Adam Niesłony, Michał Böhm, Frequency-domain fatigue life estimation with mean stress correction, International Journal of Fatigue 91 (2016) 373–381.

[76] John W. Miles, On structural fatigue under random loading, Journal of the Aeronautical Sciences 21 (11) (1954) 753–762.

[77] Igor Rychlik, On the 'narrow-band' approximation for expected fatigue damage, Probabilistic Engineering Mechanics 8 (1) (1993) 1–4.

[78] Denis Benasciutti, Robert Tovo, Rainflow cycle distribution and fatigue damage in Gaussian random loadings, Technical report, Department of Engineering, University of Ferrara, 2004.

[79] Zhen Gao, Torgeir Moan, Frequency-domain fatigue analysis of wide-band stationary Gaussian processes using a trimodal spectral formulation, International Journal of Fatigue 30 (10–11) (2008) 1944–1955.

[80] Denis Benasciutti, Roberto Tovo, Comparison of spectral methods for fatigue analysis of broad-band Gaussian random processes, Probabilistic Engineering Mechanics 21 (4) (2006) 287–299.

[81] Henrik O. Madsen, Steen Krenk, Niels C. Lind, Methods of Structural Safety, Dover Publications, 2006.

[82] Roberto Tovo, Cycle distribution and fatigue damage under broad-band random loading, International Journal of Fatigue 24 (11) (2002) 1137–1147.

[83] Matjaž Mršnik, Janko Slavič, Miha Boltežar, Multiaxial vibration fatigue—a theoretical and experimental comparison, Mechanical Systems and Signal Processing 76–77 (2016) 409–423.

[84] Loren D. Lutes, Curtis E. Larsen, Improved spectral method for variable amplitude fatigue prediction, Journal of Structural Engineering 116 (4) (1990) 1149–1164.

[85] Curtis E. Larsen, Loren D. Lutes, Predicting the fatigue life of offshore structures by the single-moment spectral method, Probabilistic Engineering Mechanics 6 (2) (1991) 96–108.

[86] Guoyang Jiao, Torgeir Moan, Probabilistic analysis of fatigue due to Gaussian load processes, Probabilistic Engineering Mechanics 5 (2) (1990) 76–83.

[87] Shinsuke Sakai, Hiroyuki Okamura, On the distribution of Rainflow range for Gaussian random processes with bimodal PSD, Transactions of the Japan Society of Mechanical Engineers Series A 60 (574) (1994) 1464–1469.

[88] Tsengti Fu, David Cebon, Predicting fatigue lives for bi-modal stress spectral densities, International Journal of Fatigue 22 (1) (2000) 11–21.

[89] Claudio Braccesi, Filippo Cianetti, Lorenzo Tomassini, Random fatigue. a new frequency domain criterion for the damage evaluation of mechanical components, International Journal of Fatigue 70 (2015) 417–427.

[90] Ed Habtour, William Connon, Michael F. Pohland, Samuel C. Stanton, Mark Paulus, Abhijit Dasgupta, Review of response and damage of linear and nonlinear systems under multiaxial vibration, Shock and Vibration 2014 (2014) 1–21.

[91] Murat Aykan, Mehmet Çelik, Vibration fatigue analysis and multi-axial effect in testing of aerospace structures, Mechanical Systems and Signal Processing 23 (3) (2009) 897–907.

[92] Xavier Pitoiset, André Preumont, Alain Kernilis, Tools for a multiaxial fatigue analysis of structures submitted to random vibrations, in: European Conference on Spacecraft Structures, Materials and Mechanical Testing, Brunswick, Germany, in: European Space Agency - Publications - ESA SP, vol. 428, 1998, pp. 289–294.

[93] Denis Benasciutti, Frank Sherratt, Alessandro Cristofori, Recent developments in frequency domain multi-axial fatigue analysis, International Journal of Fatigue 91 (2016) 397–413.

[94] Claudio Braccesi, Filippo Cianetti, Guido Lori, Dario Pioli, Random multiaxial fatigue: a comparative analysis among selected frequency and time domain fatigue evaluation methods, International Journal of Fatigue 74 (2015) 107–118.

[95] Herbert J. Gough, Engineering steels under combined cyclic and static stresses, Proceedings of the Institution of Mechanical Engineers 160 (1) (1949) 417–440.

[96] Herbert J. Gough, H.V. Pollard, The strength of metals under combined alternating stresses, Proceedings of the Institution of Mechanical Engineers 131 (1) (1935) 3–103.

[97] George Sines, Behavior of metals under complex static and alternating stresses, in: Metal Fatigue, McGraw-Hill, 1959, pp. 145–169.

[98] Benasciutti Denis, Some analytical expressions to measure the accuracy of the "equivalent von Mises stress" in vibration multiaxial fatigue, Journal of Sound and Vibration 333 (18) (2014) 4326–4340.

[99] William N. Findley, A theory for the effect of mean stress on fatigue of metals under combined torsion and axial load or bending, Journal of Engineering for Industry 81 (4) (1959) 301–306.

[100] Andrea Carpinteri, Roberto Brighenti, Ewald Macha, Andrea Spagnoli, Expected principal stress directions under multiaxial random loading. Part II: numerical simulation and experimental assessment through the weight function method, International Journal of Fatigue 21 (1) (1999) 89–96.

[101] Ioannis V. Papadopoulos, Piermaria Davoli, Carlo Gorla, Mauro Filippini, Andrea Bernasconi, A comparative study of multiaxial high-cycle fatigue criteria for metals, International Journal of Fatigue 19 (3) (1997) 219–235.

[102] Bong-Ryul You, Soon-Bok Lee, A critical review on multiaxial fatigue assessments of metals, International Journal of Fatigue 18 (4) (1996) 235–244.

[103] Xavier Pitoiset, Igor Rychlik, André Preumont, Spectral methods to estimate local multiaxial fatigue failure for structures undergoing random vibrations, Fatigue & Fracture of Engineering Materials & Structures 24 (11) (2001) 715–727.

[104] Ky Van Dang, Introduction to fatigue analysis in mechanical design by the multiscale approach, in: Ky Van Dang, Ioannis V. Papadopoulos (Eds.), High-Cycle Metal Fatigue: From Theory to Applications, in: International Centre for Mechanical Sciences, vol. 392, Springer-Verlag, 1999, pp. 57–88.

[105] Donald L. McDiarmid, A shear stress based critical-plane criterion of multiaxial fatigue failure for design and life prediction, Fatigue & Fracture of Engineering Materials & Structures 17 (12) (1994) 1475–1484.

[106] Tomokazu Matake, An explanation on fatigue limit under combined stress, Bulletin of JSME 20 (141) (1977) 257–263.

[107] Adam Niesłony, Comparison of some selected multiaxial fatigue assessment criteria, Journal of Theoretical and Applied Mechanics 48 (1) (2010) 233–254.

[108] Adam Niesłony, Ewald Macha, Spectral Method in Multiaxial Random Fatigue, Lecture Notes in Applied and Computational Mechanics, vol. 33, Springer-Verlag, 2007.

[109] André Preumont, Vincent Piéfort, Predicting random high-cycle fatigue life with finite elements, Journal of Vibration and Acoustics 116 (2) (1994) 245–248.
[110] Martijn H.A. Bonte, André D. de Boer, Ruud Liebregts, Determining the von Mises stress power spectral density for frequency domain fatigue analysis including out-of-phase stress components, Journal of Sound and Vibration 302 (1–2) (2007) 379–386.
[111] Andrea Carpinteri, Andrea Spagnoli, Multiaxial high-cycle fatigue criterion for hard metals, International Journal of Fatigue 23 (2) (2001) 135–145.
[112] Andrea Carpinteri, Andrea Spagnoli, Sabrina Vantadori, Reformulation in the frequency domain of a critical plane-based multiaxial fatigue criterion, International Journal of Fatigue 67 (2014) 55–61.
[113] Andrea Carpinteri, Giovanni Fortese, Camilla Ronchei, Daniela Scorza, Andrea Spagnoli, Sabrina Vantadori, Fatigue life evaluation of metallic structures under multiaxial random loading, International Journal of Fatigue 90 (2016) 191–199.
[114] Allan G. Davenport, Note on the distribution of the largest value of a random function with application to gust loading, in: Proceedings of the Institution of Civil Engineers, vol. 28, 1964, pp. 187–196.
[115] Alessandro Cristofori, Denis Benasciutti, Roberto Tovo, A stress invariant based spectral method to estimate fatigue life under multiaxial random loading, International Journal of Fatigue 33 (7) (2011) 887–899.
[116] Alessandro Cristofori, Luca Susmel, Roberto Tovo, A stress invariant based criterion to estimate fatigue damage under multiaxial loading, International Journal of Fatigue 30 (9) (2008) 1646–1658.
[117] Chan-Jung Kim, Yeon J. Kang, Bong-Hyun Lee, Experimental spectral damage prediction of a linear elastic system using acceleration response, Mechanical Systems and Signal Processing 25 (7) (2011) 2538–2548.
[118] Martin Česnik, Janko Slavič, Miha Boltežar, Uninterrupted and accelerated vibrational fatigue testing with simultaneous monitoring of the natural frequency and damping, Journal of Sound and Vibration 331 (24) (2012) 5370–5382.
[119] Martin Česnik, Janko Slavič, Vibrational fatigue and structural dynamics for harmonic and random loads, Journal of Mechanical Engineering 60 (5) (2014) 339–348.
[120] Claudio Braccesi, Filippo Cianetti, Lorenzo Tomassini, An innovative modal approach for frequency domain stress recovery and fatigue damage evaluation, International Journal of Fatigue 91 (2016) 382–396.
[121] Massimiliano Palmieri, Martin Česnik, Janko Slavič, Filippo Cianetti, Miha Boltežar, Non-Gaussianity and non-stationarity in vibration fatigue, International Journal of Fatigue 97 (2017) 9–19.
[122] Lorenzo Capponi, Martin Česnik, Janko Slavič, Filippo Cianetti, Miha Boltežar, Non-stationarity index in vibration fatigue: theoretical and experimental research, International Journal of Fatigue 104 (2017) 221–230.
[123] Zhengwei Fan, Yu Jiang, Shufeng Zhang, Junyong Tao, Xun Chen, Research on vibration fatigue of carbon fibre-reinforced composites under non-Gaussian random load, in: 2017 Prognostics and System Health Management Conference, PHM-Harbin, 2017.
[124] Qian Niu, Shi-xi Yang, Xing-lin Li, An empirical mode decomposition-based frequency-domain approach for the fatigue analysis of nonstationary processes, Fatigue & Fracture of Engineering Materials & Structures 41 (9) (2018) 1980–1996.
[125] Michał Böhm, Tadeusz Łagoda, Fatigue life assessment with the use of the spectral method for non-Gaussian loading histories with the use of the energy parameter, Journal of Machine Construction and Maintenance 108 (1) (2018) 27–31.
[126] Ronghui Zheng, Huaihai Chen, Xudong He, Wei Zheng, Probability distributions control for multi-input multi-output stationary non-Gaussian random vibration test, Journal of Vibration and Control 24 (21) (2017) 5201–5210.
[127] Filippo Cianetti, Massimiliano Palmieri, Claudio Braccesi, Giulia Morettini, Correction formula approach to evaluate fatigue damage induced by non-Gaussian stress state, Procedia Structural Integrity 8 (2018) 390–398.

[128] MIL-STD-810G, Test Method 514.6: Vibration, US Department of Defence, 2008.
[129] DEF STAN 00-035: PART 4, Environmental Handbook for Defence Materiel Part 4 - Natural Environments, UK Ministry of Defence, 1999.
[130] Stephen A. Rizzi, Adam Przekop, Travis Turner, On the response of a nonlinear structure to high kurtosis non-Gaussian random loadings, in: Proceedings of the 8th International Conference on Structural Dynamics (EURODYN), 2011.
[131] Frederic Kihm, Stephen A. Rizzi, Neil S. Ferguson, Andrew Halfpenny, Understanding how kurtosis is transferred from input acceleration to stress response and it's influence on fatigue life, in: Proceedings of the 11th International Conference on Recent Advances in Structural Dynamics (RASD), 2013.
[132] Adam Niesłony, Michał Böhm, Tadeusz Łagoda, Filippo Cianetti, The use of spectral method for fatigue life assessment for non-Gaussian random loads, Acta Mechanica et Automatica 10 (2) (2016) 100–103.
[133] Rainer von Sachs, Guy P. Nason, Gerald Kroisandt, Adaptive estimation of the evolutionary wavelet spectrum, Technical report, Department of Statistics, Stanford University, 1997.
[134] Wei Zhang, Chunsheng S. Cai, Fang Pan, Ye Zhang, Fatigue life estimation of existing bridges under vehicle and non-stationary hurricane wind, Journal of Wind Engineering and Industrial Aerodynamics 133 (2014) 135–145.
[135] Vincent Rouillard, Michael A. Sek, A frequency domain technique for maintaining resonance condition during sine dwell vibration testing of packages, Packaging Technology and Science 13 (6) (2000) 227–232.
[136] Heino B. Nielsen, Non-stationary time series and unit root tests, Lecture notes, 2004.
[137] Cyril Hory, Nadine Martin, Alain Chehikian, Spectrogram segmentation by means of statistical features for non-stationary signal interpretation, IEEE Transactions on Signal Processing 50 (12) (2002) 2915–2925.
[138] Jean D. Gibbons, Nonparametric Methods for Quantitative Analysis, 2nd edition, American Sciences Press, 1985.
[139] ISO 16750-3:2012, Road vehicles - Environmental conditions and testing for electrical and electronic equipment - Part 3: Mechanical loads, International Organization for Standardization, 2012.
[140] Ondej Kováík, Petr Haušild, Jií Čapek, Jan Medický, Jan Siegl, Radek Mušálek, Zdeněk Pala, Nicholas Curry, Stefan Björklund, Resonance bending fatigue testing with simultaneous damping measurement and its application on layered coatings, International Journal of Fatigue 82 (2016) 300–309.
[141] Robert Owsiński, Adam Niesłony, Review of current state of knowledge on durability tests performed on electromagnetic shakers, Modelling in Engineering 22 (53) (2014) 117–123.
[142] Leila Khalij, Christophe Gautrelet, Alain Guillet, Fatigue curves of a low carbon steel obtained from vibration experiments with an electrodynamic shaker, Materials & Design 86 (2015) 640–648.
[143] Yu Jiang, Gun Jin Yun, L Zhao, Junyong Tao, Experimental design and validation of an accelerated random vibration fatigue testing methodology, Shock and Vibration 2015 (2015).
[144] Ondej Kováík, Petr Haušild, Jan Medický, Libor Tomek, Jan Siegl, Radek Mušálek, Nicholas Curry, Stefan Björklund, Fatigue crack growth in bodies with thermally sprayed coating, Journal of Thermal Spray Technology 25 (1–2) (2016) 311–320.
[145] Kishor K. Meher, J.K. Pandey, A. RamaRao, Accelerated vibration test of coolant channel components under simulated flow induced excitation, Nuclear Engineering and Design 300 (2016) 127–134.
[146] Uroš Proso, Janko Slavič, Miha Boltežar, Vibration-fatigue damage accumulation for structural dynamics with non-linearities, International Journal of Mechanical Sciences 106 (2016) 72–77.
[147] Vanja Pahor Kos, Janko Slavi, Miha Boltear, Fatigue damage for sweep-sine and random accelerated vibration testing, Advances in Mechanical Engineering 7 (2) (2015) 340545.

[148] Toumi M. Yessine, Fabrice Bolaers, Fabien Bogard, Sébastien Murer, Study of ball bearing fatigue damage using vibration analysis: application to thrust ball bearings, Structural Engineering and Mechanics 53 (2) (2015) 325–336.
[149] Matija Brumat, Janko Slavič, Miha Boltežar, Design of damping layout using spatial-damping identification methods, International Journal of Mechanical Sciences 127 (2017) 41–46.
[150] Tommy J. George, Jeremy Seidt, Mo-How Herman Shen, Theodore Nicholas, Charles J. Cross, Development of a novel vibration-based fatigue testing methodology, International Journal of Fatigue 26 (5) (2004) 477–486.
[151] Bernhard Kurth, Hubert Bomas, Peter Mayr, Einflußder Oberflächenbeschaffenheit auf die Schwingfestigkeit von Aluminium-Druckguß, Materialwissenschaft und Werkstofftechnik 26 (2) (1995) 101–110.
[152] John G. Kaufman, Properties of Aluminum Alloys: Fatigue Data and the Effects of Temperature, Product Form, and Processing, ASM International, 2008.
[153] Massimiliano Avalle, Giovanni Belingardi, Maria P. Cavatorta, Static and fatigue strength of a die cast aluminium alloy under different feeding conditions, Metallurgical Science and Technology 20 (2) (2002) 9–13.
[154] Lech Dietrich, Joanna Radziejewska, The fatigue damage development in a cast Al-Si-Cu alloy, Materials & Design 32 (1) (2011) 322–329.
[155] Rodrigo González, Dora I. Martínez, J. Alejandro González, José Talamantes, Salvador Valtierra, Rafael Colás, Experimental investigation for fatigue strength of a cast aluminium alloy, International Journal of Fatigue 33 (2) (2011) 273–278.
[156] Mohd A.M. Daud, Mohd Z. Omar, Junaidi Syarif, Zainuddin Sajuri, Effect of wire-EDM cutting on fatigue strength of AZ61 magnesium alloy, Jurnal Mekanikal 30 (30) (2010) 68–76.
[157] Rui-Jie Wang, De-Guang Shang, Fatigue life prediction based on natural frequency changes for spot welds under random loading, International Journal of Fatigue 31 (2) (2009) 361–366.
[158] Curtis E. Larsen, Tom Irvine, A review of spectral methods for variable amplitude fatigue prediction and new results, Procedia Engineering 101 (2015) 243–250.
[159] Neil W.M. Bishop, Z. Hu, Rui Wang, David Quarton, Methods for rapid evaluation of fatigue damage on the Howden HWP 330 wind turbine, in: Proceedings of 15th Wind Energy Conference, Wind Energy Conversion, York, vol. 1, 1993, pp. 261–268.
[160] Vincent Bouyssy, S.M. Naboishikov, Ruediger Rackwitz, Comparison of analytical counting methods for Gaussian processes, Structural Safety 12 (1) (1993) 35–57.
[161] Giovanni Petrucci, Bernardo Zuccarello, On the estimation of the fatigue cycle distribution from spectral density data, Proceedings of the Institution of Mechanical Engineers, Part C: Journal of Mechanical Engineering Science 213 (8) (1999) 819–831.
[162] Tadeusz Łagoda, Ewald Macha, Adam Niesłony, Fatigue life calculation by means of the cycle counting and spectral methods under multiaxial random loading, Fatigue & Fracture of Engineering Materials & Structures 28 (4) (2005) 409–420.
[163] Claudio Braccesi, Filippo Cianetti, Guido Lori, Dario Pioli, Fatigue behaviour analysis of mechanical components subject to random bimodal stress process: frequency domain approach, International Journal of Fatigue 27 (4) (2005) 335–345.
[164] Giovanni Petrucci, Bernardo Zuccarello, Fatigue life prediction under wide band random loading, Fatigue & Fracture of Engineering Materials & Structures 27 (12) (2004) 1183–1195.
[165] Jeong-Tae Kim, Yeon-Sun Ryu, Hyun-Man Cho, Norris Stubbs, Damage identification in beam-type structures: frequency-based method vs mode-shape-based method, Engineering Structures 25 (1) (2003) 57–67.
[166] Ciyou Zhu, Richard H. Byrd, Peihuang Lu, Jorge Nocedal, Algorithm 778: L-BFGS-B: Fortran subroutines for large-scale bound-constrained optimization, ACM Transactions on Mathematical Software (TOMS) 23 (4) (1997) 550–560.
[167] Marta Kurek, Tadeusz Łagoda, Comparison of the fatigue characteristics for some selected structural materials under bending and torsion, Materials Science 47 (3) (2011) 334–344.
[168] Richard L. Burden, J. Douglas Faires, Numerical Analysis, 8th edition, Thomson Brooks/Cole, 2005.

Index

D

E

F